Recognizing Outstanding Ph.D. Research

Aims and Scope

The series "Springer Theses" brings together a selection of the very best Ph.D. theses from around the world and across the physical sciences. Nominated and endorsed by two recognized specialists, each published volume has been selected for its scientific excellence and the high impact of its contents for the pertinent field of research. For greater accessibility to non-specialists, the published versions include an extended introduction, as well as a foreword by the student's supervisor explaining the special relevance of the work for the field. As a whole, the series will provide a valuable resource both for newcomers to the research fields described, and for other scientists seeking detailed background information on special questions. Finally, it provides an accredited documentation of the valuable contributions made by today's younger generation of scientists.

Theses are accepted into the series by invited nomination only and must fulfill all of the following criteria

- They must be written in good English.
- The topic should fall within the confines of Chemistry, Physics, Earth Sciences, Engineering and related interdisciplinary fields such as Materials, Nanoscience, Chemical Engineering, Complex Systems and Biophysics.
- The work reported in the thesis must represent a significant scientific advance.
- If the thesis includes previously published material, permission to reproduce this must be gained from the respective copyright holder.
- They must have been examined and passed during the 12 months prior to nomination.
- Each thesis should include a foreword by the supervisor outlining the significance of its content.
- The theses should have a clearly defined structure including an introduction accessible to scientists not expert in that particular field.

Laura Ratcliff

Optical Absorption Spectra Calculated Using Linear-Scaling Density-Functional Theory

Doctoral Thesis accepted by
Imperial College London, London, UK

 Springer

Author
Dr. Laura Ratcliff
Laboratoire de simulation Atomistique
UMR-E CEA/UJF-Grenoble 1
Grenoble
France

Supervisor
Prof. Peter Haynes
Departments of Physics and Materials
Imperial College London
London
UK

ISSN 2190-5053
ISBN 978-3-319-03373-0
DOI 10.1007/978-3-319-00339-9
Springer Cham Heidelberg New York Dordrecht London

ISSN 2190-5061 (electronic)
ISBN 978-3-319-00339-9 (eBook)

Supervisor's Foreword

The digital computer has revolutionised the way science is done, creating a third dimension for scientific inquiry alongside theory and experiment in the form of simulation. Nowhere is this more evident than in materials science. In particular, the growing impact of first-principles approaches to simulations at the level of individual atoms has been truly remarkable. This class of methods starts from scratch with the equations of quantum mechanics that describe the behaviour of the electrons responsible for chemical bonds and solves them using a series of controlled approximations. Because the only knowledge assumed about the material to be studied is its chemical composition, the results are free from prior assumptions and the method can be applied to all classes of materials. Furthermore, the power of first-principles simulations becomes apparent when they are used to study the behaviour of materials under external conditions of temperature and pressure that cannot be achieved in the laboratory or even to study materials that do not (yet) exist. This is only possible because of the parameter-free approach based on the theory of quantum mechanics that underlies all of low-energy physics, chemistry and biology.

First-principles simulations owe their success to the development of Density-Functional Theory (DFT) in the 1960s. In 1998, Walter Kohn was awarded the Nobel Prize for Chemistry for his work on DFT. The dominance of DFT today is the result of a balancing act: on the one hand it provides a sufficiently accurate description of the correlated motion of electrons in materials for most purposes, on the other the computational cost of these simulations is relatively low compared to other first-principles methods. In particular, as the number of atoms $\mathcal{N}$ in the system increases, the workload for a traditional DFT simulation grows only as $\mathcal{N}^3$. This cubic scaling has allowed simulations of hundreds of atoms to be performed on a routine basis. However, ultimately it places a constraint on the length-scale accessible with DFT, and hence over the last 20 years there has been considerable work on the formulation and implementation of so-called linear-scaling methods, where the computational effort increases in direct proportion to $\mathcal{N}$. A number of software packages implementing these methods now exist, and have been used to perform simulations of many thousands of atoms. Entire nanostructures and protein molecules have been brought within the scope of DFT simulations, and new opportunities have opened up as a result of the overlap of the length-scales accessible to first-principles simulations and experimental characterisation.

However the development of practical schemes for exploiting linear-scaling DFT has brought new challenges that do not concern the availability of computational resources or the technical implementation of new methodology, but rather the human imagination. The issue is that as the size of accessible systems increases and more realistic, complex models of materials become available, so the number of possible configurations of those systems explodes. With traditional DFT methods the simulations tend to be constrained by size, which means that all relevant arrangements may be studied exhaustively. However as the number of atoms increases with the use of linear-scaling methods, there is an exponential growth in the number of combinations to study. New strategies are required to explore the phase space of possible structures. At the same time, computation of the total energy and forces (the standard outputs from first-principles simulations) is a rather modest return on the investment in a large-scale simulation, which has involved the calculation of details of the electronic structure. There is surely more information to extract.

The thesis work by Laura Ratcliff described in this volume aims to address both of these problems by enabling the calculation of optical absorption spectra within linear-scaling DFT simulations. This is the first example of a more general strategy to link simulation more closely with experiment by direct calculation of those properties measured in the laboratory. This extracts relevant detailed information from the calculations, but also enables the flexibility and control of simulation to be used to guide the search for relevant structures. For example, changes to the simulated structure can be made to observe their effect on the calculated spectrum and its agreement with experiment, or the spectrum may be decomposed spatially to aid association with a particular structural motif. Although a wide range of experimental spectra can now be calculated within traditional DFT, this work represents the first time this has been achieved within a fully linear-scaling formulation. Along the way, significant technical challenges must be overcome, not least the difficulty of representing excited electronic states with the spatially localised orbitals required for linear scaling. But it is work like this on the development of new functionality within software packages such as ONETEP that is essential if the revolution in the scope and scale of first-principles simulations that has taken place over the last 50 years is to continue.

London, January 2013   Peter Haynes

Acknowledgments

Can you fathom the mysteries of God?

Can you probe the limits of the Almighty?

They are higher than the heavens above—what can you do?

They are deeper than the depths below—what can you know?

Their measure is longer than the earth

and wider than the sea.

Job 11:7–9

First of all, I would like to thank my supervisor, Peter Haynes, who provided the original inspiration for this work, and without whom it would not have been possible. I am greatly indebted to you for your wisdom, support and guidance, and not least for reading through this thesis. Thank you also to Nick Hine, in particular for your help with the implementation in ONETEP; without your efforts this work would have taken considerably longer, and the final code would be far less elegant.

Another group of people to whom I am incredibly grateful are all my friends at St Nicks. Thank you for your friendship, encouragement and interest in my research throughout the past 3 years, your support has been particularly invaluable in tough times.

To Jessica Lloyd, for being one of the best housemates anyone could wish for, and to Michelle Rogers, for being our honorary housemate. The last 3, or rather 14, years have been made far more fun by having you around and I'm honestly not sure what I would have done without you. I may be leaving London soon but don't think you'll be able to escape me, I shall always be in need of random road trips and friends to sing along in the car with!

Completing this thesis would have been much harder without the support of my family, even if the closest any of them has come to understanding my work is in postulating a theory relating electrons and cows, and even then when trapped on a boat in Norway with no other option but to listen to me talk about my research!

And finally, to Fabiano Corsetti and Jo Sarsam, for all the time we've spent (wasted?) together during our PhDs, watching The Big Bang Theory/Numb3rs, playing Bejewelled/Chromatron/Peggle, eating ice cream (on a Tuesday), tampering with each other's computers when we weren't looking, solving crosswords/ the dreaded enigmas, attempting to solve Bloxx, taking endless photos and all the other hilarity I may or may not have forgotten about! As a result of your friendship I'm now considerably more geeky than I was a few years ago. Thanks also for putting up with me at my most tired, although I'm sure by now I must have paid you back in cakes, cookies and cushions! Here's to many more years of friendship (by which I of course mean ultra-serious scientific collaborations).

This work was supported by the UK Engineering and Physical Sciences Research Council (EPSRC), to whom I'm duly grateful and calculations were performed on CX1 (Imperial College London High Performance Computing Service). There are many other people I would like to thank for their help in getting me through these last 3 years, although there are too many to mention all by name. If I've not included you, sorry, it doesn't mean you're not appreciated!

Contents

Chapter 1
Introduction

The field of experimental spectroscopy involves the study of interactions between radiation and matter, and so can be very useful in probing the structure and properties of different materials. It encompasses a wide range of techniques covering all regions of the electromagnetic spectrum, or in the case of electron based techniques such as electron energy loss spectroscopy (EELS), a wide range of electron energies. It can be used to probe bulk systems, molecules and surfaces. High resolutions and short measurement time scales can now be achieved, allowing for example the study of transient structures in chemical reactions [1] and the movement of atoms at surfaces [2].

The calculation of experimental spectra is an important tool, and has been the focus of much research. One of the chief benefits of theoretical spectroscopy is that it allows the prediction of experimental results for new and existing materials. More significantly, it can also be used to further our understanding of spectra, for example by identifying the electronic transitions responsible for a given peak. The comparison of theoretical and experimental results can also be used to distinguish between candidate structures for a material [3, 4]. With these being just some of the uses for the field of theoretical spectroscopy, it is therefore desirable to have an efficient and reliable *ab initio* method for calculating spectra. That is, we would like to be able to calculate such spectra without needing any prior knowledge of the material in question.

Density-functional theory (DFT) [5, 6] is a highly successful formulation of quantum mechanics which has allowed the study of a wide variety of electronic systems and their material properties. A number of DFT codes have been specifically designed to take advantage of the ever increasing capabilities of modern computers, so that large scale parallel calculations can now be routinely performed for systems containing up to several hundred atoms. This advance in *ab initio* simulations has resulted in the field of computational materials simulation becoming fundamental in advancing our understanding of current materials. Furthermore, it can be used to help improve the properties of current materials, as well as develop new ones with specific applications in mind [7, 8].

L. Ratcliff, *Optical Absorption Spectra Calculated Using Linear-Scaling Density-Functional Theory*, Springer Theses, DOI: 10.1007/978-3-319-00339-9_1,

DFT is therefore a good initial framework in which to calculate the energy eigenstates which are required for the calculation of such spectra. In practice, however, many systems of interest are large in scale, and as such computationally expensive, if not impossible, to treat with traditional approaches to DFT, where the computational effort scales as the cube of the system size. However, DFT can also be reformulated to scale only linearly with system size, which requires the use of local orbitals [9–13]. This offers the opportunity to access much larger system sizes, and if combined with theoretical spectroscopy, it could become a very powerful tool. The combination of the two methods is particularly important in that it allows one to extract information from calculations on large systems that can be directly compared to measurable properties of the system.

There is, however, a substantial challenge in using the results of local orbital calculations to generate spectra, in that the orbitals are optimized to describe the occupied states, which results in a fundamental problem in the representation of unoccupied states. There are two approaches to the optimization of such orbitals; either via the use of basis sets of purpose-designed atomic orbitals, or via the minimization of total energy with respect to some set of local orbitals which therefore become adapted to the system in question, which is the approach followed in this work. In both cases, this results in a basis which is unable to represent the unoccupied states very well. This problem is particularly noticeable in systematically convergeable linear-scaling methods such as ONETEP [14–17], where very precise agreement can be reached with traditional DFT codes for the occupied states, whilst the unoccupied states differ significantly, with some states being completely absent [18].

In order to overcome the problems with representing unoccupied states, we have developed a new method whereby a second set of localized functions is optimized specifically to describe the unoccupied states. With this method, it becomes possible to implement the calculation of optical absorption spectra using Fermi's golden rule. It could also be extended to other types of spectroscopy in future, such as EELS.

Within this thesis we will present our approach for the calculation of optical absorption spectra within a linear-scaling DFT framework. This has been applied in ONETEP and so this is the method on which we will focus but it is equally applicable to other local-orbital methods. We first begin by outlining DFT and its application to periodic systems in Chap. 2. We continue in Chap. 3 by presenting an introduction to linear-scaling formalisms of DFT, including a survey of the different approaches to achieving linear-scaling and a more detailed discussion of the theory behind the ONETEP code. In Chap. 4 we then give an overview of the field of theoretical spectroscopy, focussing in more detail on the method we shall be using to calculate optical absorption spectra. In Chap. 5 we introduce a "toy model" which we have used to compare different methods for the calculation of conduction states. This model includes the use of a localized basis set, and so we also compare some simple one-dimensional localized basis sets, identifying some of the features which are desirable for such basis sets, which can be related to the local orbitals used in ONETEP. Furthermore, we discuss two different methods for the calculation of band structures within localized basis sets. Chapter 6 contains a summary of the different methods for calculating conduction states, as well as the results from the toy model which were used to select the best method. This method is then further developed and the details for the implementation in ONETEP are given. In Chap. 7 we present our results for both molecular systems and extended polymers. These results have been validated through comparison with results from a traditional cubic-scaling DFT code; excellent agreement between the two is demonstrated. Finally in Chap. 8 we present our conclusions and identify some areas for future work.

References

1. Ihee, H., Lobastov, V.A., Gomez, U.M., Goodson, B.M., et al.: Science **291**(5503), 458 (2001)
2. Petek, H., Weida, M.J., Nagano, H., Ogawa, S.: Science **288**(5470), 1402 (2000)
3. Kibalchenko, M., Lee, D., Shao, L., Payne, M.C., et al.: Chem. Phys. Lett. **498**(4–6), 270 (2010)
4. Kibalchenko, M., Yates, J.R., Massobrio, C., Pasquarello, A.: Phys. Rev. B **82**, 020202 (2010)
5. Hohenberg, P., Kohn, W.: Phys. Rev. **136**(3B), B864 (1964)
6. Kohn, W., Sham, L.J.: Phys. Rev. **140**(4A), A1133 (1965)
7. Marzari, N.: MRS Bull. **31**, 681 (2006)
8. Hafner, J., Wolverton, C., Ceder, G.: MRS Bull. **31**, 659 (2006)
9. Galli, G., Parrinello, M.: Phys. Rev. Lett. **69**(24), 3547 (1992)
10. Ordejón, P., Drabold, D.A., Martin, R.M., Grumbach, M.P.: Phys. Rev. B **51**(3), 1456 (1995)
11. Hernández, E., Gillan, M.J.: Phys. Rev. B **51**(15), 10157 (1995)
12. Fattebert, J.-L., Bernholc, J.: Phys. Rev. B **62**(3), 1713 (2000)
13. Skylaris, C.-K., Mostofi, A.A., Haynes, P.D., Diéguez, O., et al.: Phys. Rev. B **66**(3), 035119 (2002)
14. Skylaris, C.-K., Haynes, P.D., Mostofi, A.A., Payne, M.C.: J. Chem. Phys. **122**(8), 084119 (2005)
15. Haynes, P.D., Skylaris, C.-K., Mostofi, A.A., Payne, M.C.: Phys. Status Solidi B **243**(11), 2489 (2006)
16. Hine, N.D.M., Haynes, P.D., Mostofi, A.A., Skylaris, C.-K., et al.: Comput. Phys. Commun. **180**(7), 1041 (2009)
17. Hine, N.D.M., Robinson, M., Haynes, P.D., Skylaris, C.-K., et al.: Phys. Rev. B **83**(19), 195102 (2011)
18. Skylaris, C.-K., Haynes, P.D., Mostofi, A.A., Payne, M.C.: J. Phys.: Condens. Matter **17**(37), 5757 (2005)

Chapter 2
Density-Functional Theory

In this chapter we introduce the problem of interacting electronic systems, and in particular our method of choice, density-functional theory. We first present the full many-body Hamiltonian and briefly outline the Born-Oppenheimer approximation. Following an introduction to DFT we discuss some of the concepts required for the application to periodic systems, namely supercells, Bloch's theorem, plane-waves and pseudopotentials.

Whilst there are other popular and successful methods for treating electronic systems quantum mechanically, including for example Hartree-Fock theory, we do not discuss these approaches, instead limiting ourselves to DFT. The interested reader may refer to e.g. [1–3] for further information on such methods, as well as a more detailed introduction to DFT than that provided here.

2.1 Many-Electron Systems

Starting from the principles of quantum mechanics, it should, in theory, be possible to determine any physical property of a system of many particles. Indeed, it is possible to write down the time-dependent Schrödinger equation (TDSE), as:

$$\hat{\mathcal{H}}|\Psi\rangle = i\frac{\partial|\Psi\rangle}{\partial t}, \tag{2.1}$$

where Dirac notation has been employed to write the wavefunction $|\Psi\rangle$ and $\hat{\mathcal{H}}$ is the Hamiltonian, or energy operator, defined as:

$$\hat{\mathcal{H}} = -\frac{1}{2}\sum_i \nabla_i^2 - \sum_\alpha \frac{1}{2M_\alpha}\nabla_\alpha^2 - \sum_i \sum_\alpha \frac{Z_\alpha}{|\mathbf{r}_i - \mathbf{R}_\alpha|}$$
$$+ \frac{1}{2}\sum_i \sum_{j \neq i} \frac{1}{|\mathbf{r}_i - \mathbf{r}_j|} + \frac{1}{2}\sum_\alpha \sum_{\beta \neq \alpha} \frac{Z_\alpha Z_\beta}{|\mathbf{R}_\alpha - \mathbf{R}_\beta|} \tag{2.2}$$

L. Ratcliff, *Optical Absorption Spectra Calculated Using Linear-Scaling Density-Functional Theory*, Springer Theses, DOI: 10.1007/978-3-319-00339-9_2,
© Springer International Publishing Switzerland 2013

for a system of interacting electrons and nuclei, where we have neglected relativistic effects. Greek indices refer to nuclei and Latin indices are used for electrons, so that the $\mathbf{r}_i$ are the electronic positions and $\mathbf{R}_\alpha$ are the nuclear coordinates. The Z_α are atomic numbers and the M_α are the nuclear masses. Hartree atomic units are used here and throughout this thesis unless otherwise stated, such that $\hbar = m_e = e = 4\pi\varepsilon_0 = 1$. For the purposes of this work, spin will be neglected, however all equations can be generalized to include its effect e.g. spin-polarized systems.

However, the complexity of the above Schrödinger equation and the extremely large number of variables involved in the problem make it impossible to solve directly for all but the simplest of systems, either analytically or numerically. Thus it becomes necessary to use approximate methods. We first proceed by performing a separation of variables, so that the wavefunction is written as:

$$\Psi\left(\{\mathbf{r}_i\}, \{\mathbf{R}_\alpha\}, t\right) = \Phi\left(\{\mathbf{r}_i\}, \{\mathbf{R}_\alpha\}\right)\Theta\left(t\right). \tag{2.3}$$

Substituting this into the TDSE (Eq. 2.1), we derive the following two equations:

$$\hat{\mathcal{H}}\Phi\left(\{\mathbf{r}_i\}, \{\mathbf{R}_\alpha\}\right) = \mathcal{E}\Phi\left(\{\mathbf{r}_i\}, \{\mathbf{R}_\alpha\}\right) \tag{2.4}$$

$$\mathrm{i}\frac{\mathrm{d}\Theta\left(t\right)}{\mathrm{d}t} = \mathcal{E}\Theta\left(t\right), \tag{2.5}$$

where $\mathcal{E}$ is the energy of a given eigenstate and the first equation is the time-independent Schrödinger equation (TISE). Equation 2.5 can now be solved so that the overall wavefunction becomes:

$$\Psi\left(\{\mathbf{r}_i\}, \{\mathbf{R}_\alpha\}, t\right) = \Phi\left(\{\mathbf{r}_i\}, \{\mathbf{R}_\alpha\}\right)\mathrm{e}^{-\mathrm{i}\mathcal{E}t}. \tag{2.6}$$

Having separated out the time-dependent component, we can now continue by focussing on how to solve for the stationary part of the wavefunction only, i.e. $\Phi\left(\{\mathbf{r}_i\}, \{\mathbf{R}_\alpha\}\right)$, rather than the full wavefunction $\Psi\left(\{\mathbf{r}_i\}, \{\mathbf{R}_\alpha\}, t\right)$.

One can then proceed by considering the relative time scales of nuclear versus electronic motion; as the nuclear masses are so much larger than the electronic masses, whilst their momenta will be of the same order of magnitude, it is reasonable to make the approximation that the electronic velocities will be considerably greater than the nuclear velocities. Therefore the electronic motion can be treated as effectively instantaneous compared to the nuclear movement. This allows us to make a further separation of variables, such that for a given configuration, the nuclei are considered to be stationary and so the electronic ground state energy depends only parametrically on the nuclear coordinates. The wavefunction can therefore be written in the following approximate form:

$$\Phi\left(\{\mathbf{r}_i\}, \{\mathbf{R}_\alpha\}\right) = \psi\left(\{\mathbf{r}_i\}; \{\mathbf{R}_\alpha\}\right)\zeta\left(\mathbf{R}_\alpha\right) \tag{2.7}$$

and the new TISE for the electronic wavefunction is:

$$\hat{H}\psi\left(\{\mathbf{r}_i\};\{\mathbf{R}_\alpha\}\right) = E\psi\left(\{\mathbf{r}_i\};\{\mathbf{R}_\alpha\}\right),\tag{2.8}$$

where the electronic Hamiltonian, $\hat{H}$ is:

$$\hat{H} = -\frac{1}{2}\sum_i \nabla_i^2 - \sum_i \sum_\alpha \frac{Z_\alpha}{|\mathbf{r}_i - \mathbf{R}_\alpha|} + \frac{1}{2}\sum_i \sum_{j\neq i} \frac{1}{|\mathbf{r}_i - \mathbf{r}_j|}.\tag{2.9}$$

This approach is known as the Born-Oppenheimer approximation [4].

For small systems with very few electrons, it is possible to solve Eq. (2.8) directly, using wavefunction methods involving greater or fewer approximations. However, for bigger systems, essentially exact methods such as the full configuration interaction method scale exponentially, and even more approximate correlated wavefunction methods scale as $\mathcal{O}\left(\mathcal{N}^5\right)$-$\mathcal{O}\left(\mathcal{N}^7\right)$. This results in a strict limit on the number of electrons which can be treated using such methods. We therefore find ourselves seeking an alternative approach to the quantum mechanical treatment of large electronic systems, which comes in the form of DFT.

2.2 Density-Functional Theory

Density-functional theory is a well-established formalism of quantum mechanics which is based on electronic charge densities, rather than wavefunctions, thus greatly reducing the number of variables of the central quantity from $3N$ to 3, where N is the number of electrons in the system. The earliest form of DFT was initially proposed by Thomas [5] and Fermi [6] in 1927, but was largely overlooked due to serious deficiencies in their approximations. In 1964 Hohenberg and Kohn [7] introduced proofs for their two theorems which demonstrate the fundamental importance of the electronic density, and in 1965 Kohn and Sham [8] further developed the theory into a framework for real calculations. Today DFT is widely used by physicists and chemists alike, and its importance was recognised with the award of the 1998 Nobel prize for Chemistry to Walter Kohn and John Pople.

2.2.1 Hohenberg and Kohn: The Two Theorems

Thomas and Fermi proposed a method for calculating the energy of a system solely in terms of the electronic density. The simplifications involved in their approach restricted its usefulness, so that it was not quantitatively applicable for the study of most systems. However, the idea of using the electronic density as the central variable proved to be the inspiration for DFT.

Hohenberg and Kohn outlined DFT in terms of two theorems, proving the existence of a one-to-one mapping between the potential $V(\mathbf{r})$ and the density $n(\mathbf{r})$. The first of these theorems is as follows, where the proof is for a non-degenerate ground state.

Theorem 1 *The ground state density of a system of interacting electrons in an external potential uniquely defines this potential, except for an additive constant.*

Proof We start by assuming the existence of two separate potentials, $V(\mathbf{r})$ and $V'(\mathbf{r})$, which differ by more than an additive constant, and both give the same ground state density $n(\mathbf{r})$. For each potential there will also be a corresponding Hamiltonian, $\hat{H}$ and $\hat{H}'$, and a ground state wavefunction, $|\psi\rangle$ and $|\psi'\rangle$. As $|\psi'\rangle$ is not a ground state of $\hat{H}$, we can write the energy E as:

$$E = \langle\psi|\hat{H}|\psi\rangle < \langle\psi'|\hat{H}|\psi'\rangle, \tag{2.10}$$

where we make use of the variational principle. We then rewrite the last term in Eq. (2.10) as:

$$\langle\psi'|\hat{H}|\psi'\rangle = \langle\psi'|\hat{H}'|\psi'\rangle + \langle\psi'|\left(\hat{H} - \hat{H}'\right)|\psi'\rangle \tag{2.11}$$

$$= E' + \int \left[V(\mathbf{r}) - V'(\mathbf{r})\right] n(\mathbf{r})\, d\mathbf{r}, \tag{2.12}$$

where E' is the energy of the second system. Substituting back into Eq. (2.10), we have:

$$E < E' + \int \left[V(\mathbf{r}) - V'(\mathbf{r})\right] n(\mathbf{r})\, d\mathbf{r}. \tag{2.13}$$

Following a similar procedure, we could also have written:

$$E' < E + \int \left[V'(\mathbf{r}) - V(\mathbf{r})\right] n(\mathbf{r})\, d\mathbf{r}. \tag{2.14}$$

Adding Eqs. (2.13) and (2.14), we arrive at:

$$E + E' < E' + E, \tag{2.15}$$

which is clearly a contradiction. By *reductio ad absurdum* it therefore follows that there can be no two different potentials giving rise to the same ground state density, thus proving the theorem.

It also follows on from this theorem that as the Hamiltonian and total number of electrons, N, is also uniquely determined by the density, so too will the many-body wavefunctions and thus all other properties of the system which stem from the Hamiltonian.

The second theorem concerns the existence of a universal functional of the total energy in terms of the density, and is defined as follows.

Theorem 2 *There exists a universal functional for the energy in terms of the density, which is valid for any external potential. For a given external potential, the* *global minimum of this functional is the ground state energy of the system, and the corresponding density is the ground state density.*

Proof We first observe that as the wavefunction is a functional of the density, so too are the kinetic and interaction energies, so that the total energy functional can be written as:

$$E_V[n] = \int V(\mathbf{r})\, n(\mathbf{r})\, d\mathbf{r} + F[n], \tag{2.16}$$

where $F[n]$ is defined as the sum of the kinetic, $T[n]$, and interaction, $U[n]$, energies, i.e.:

$$F[n] = T[n] + U[n]. \tag{2.17}$$

$F[n]$ is a *universal* functional valid for any number of electrons with any external potential, as the kinetic and interaction energies depend only on the density. If we now take a system with a ground state density $n(\mathbf{r})$ and external potential $V(\mathbf{r})$, we can equate the ground state energy E with the expectation value of the Hamiltonian $\hat{H}$, via the wavefunction $|\psi\rangle$, and the energy functional defined in Eq. (2.16):

$$E = E_V[n] = \langle\psi|\hat{H}|\psi\rangle. \tag{2.18}$$

If we now take a different density $n'(\mathbf{r})$ (which is the ground state density of a different external potential $V'(\mathbf{r})$, a property which is known as V-representability), which we shall refer to as the trial density, and its corresponding wavefunction $|\psi'\rangle$ and energy E', we can use the variational principle to state that:

$$E = \langle\psi|\hat{H}|\psi\rangle < \langle\psi'|\hat{H}|\psi'\rangle = E', \tag{2.19}$$

or in other words:

$$E_V\left[n'\right] > E_V[n]. \tag{2.20}$$

Therefore the energy is indeed at a minimum for the ground state density, and if the universal functional could be found, then one could minimize the total energy as defined by the functional, and thus obtain the ground state density.

Levy [9, 10] and Lieb [11, 12] have provided an alternative, more general, formulation of DFT known as the constrained search formulation. Their redefinition of the universal functional allows for the restriction to non-degenerate ground states to be lifted and widens the range of allowed trial densities to include those which are N-representable, i.e. can be derived from an anti-symmetric N-body wavefunction, rather than just those which are V-representable. A number of extensions of DFT also exist, including spin DFT, ensemble DFT and current-DFT, for more information on which the interested reader is referred to e.g. [1].

2.2.2 The Kohn–Sham Equations

Hohenberg and Kohn proceeded by rewriting their energy functional to include the classical electrostatic, or Hartree term, giving:

$$E_V[n] = \int V(\mathbf{r})\, n(\mathbf{r})\, d\mathbf{r} + \frac{1}{2} \iint \frac{n(\mathbf{r})\, n(\mathbf{r}')}{|\mathbf{r} - \mathbf{r}'|}\, d\mathbf{r}\, d\mathbf{r}' + G[n], \tag{2.21}$$

where $G[n]$, like $F[n]$, is a universal functional, i.e. its definition is independent of $V(\mathbf{r})$. The problem then remained of finding an expression for $G[n]$ which would result in a usable scheme for actually calculating the ground state energy. In order to achieve this goal, Kohn and Sham made the novel step of introducing an auxiliary system of non-interacting "electrons", where the two systems have the same density. This process divides the unknown functional into separate components, whereby those components whose definition is unknown are expected to be smaller in magnitude, and so approximations can be made with the hope of retaining reasonable accuracy in the final solutions. In particular, this step allows for the definition of the kinetic energy component, which is unknown for an interacting system but can easily be defined for a non-interacting system. Whilst this will not give the true kinetic energy of the interacting system, the differences are expected to be reasonably small.

The functional $G[n]$ is therefore defined as follows:

$$G[n] = T_s[n] + E_{xc}[n], \tag{2.22}$$

where $T_s[n]$ is the kinetic energy of the auxiliary system and $E_{xc}[n]$ is defined as the exchange and correlation energy of the interacting system. It is this term within which the remaining many-body components of the energy are absorbed, as well as the difference between the interacting and non-interacting kinetic energies. Its exact form remains unknown and so various approximations must be used, which are discussed further in Sect. 2.2.3.

The auxiliary system is set up to have the same density as the original system via the use of an effective potential, $V_{eff}(\mathbf{r})$, as follows. We can use the Hohenberg-Kohn variational principle to write:

$$\delta \left\{ F[n] + \int V(\mathbf{r})\, n(\mathbf{r})\, d\mathbf{r} - \mu \left[\int n(\mathbf{r})\, d\mathbf{r} - N \right] \right\} = 0, \tag{2.23}$$

where we have introduced a Lagrange multiplier, μ, to ensure the density conserves the correct total electron number N. Using our previous definition of the functional $F[n]$, we can also write Eq. (2.23) as:

$$\frac{\delta T_s[n]}{\delta n(\mathbf{r})} + V_{eff}(\mathbf{r}) = \mu, \tag{2.24}$$

defining the effective potential $V_{\text{eff}}(\mathbf{r})$ as:

$$V_{\text{eff}}(\mathbf{r}) = V(\mathbf{r}) + V_{\text{H}}(\mathbf{r}) + V_{\text{xc}}(\mathbf{r}), \tag{2.25}$$

where $V_{\text{xc}}(\mathbf{r})$ is the exchange-correlation potential, found from $E_{\text{xc}}[n]$ via:

$$V_{\text{xc}}(\mathbf{r}) = \frac{\delta E_{\text{xc}}[n]}{\delta n(\mathbf{r})}, \tag{2.26}$$

the Hartree potential $V_{\text{H}}(\mathbf{r})$ is defined as:

$$V_{\text{H}}(\mathbf{r}) = \int \frac{n(\mathbf{r}')}{|\mathbf{r} - \mathbf{r}'|} d\mathbf{r}', \tag{2.27}$$

and $V(\mathbf{r})$ is the usual external potential.

Equation (2.24) is exactly what would have been obtained for a system of non-interacting particles in an external potential $V_{\text{eff}}(\mathbf{r})$, and so finding the ground state density which satisfies Eq. (2.24) can be achieved by solving the following set of single-particle equations:

$$\left[-\frac{1}{2}\nabla^2 + V_{\text{eff}}(\mathbf{r}) \right] |\phi_n\rangle = \mathcal{E}_n |\phi_n\rangle, \tag{2.28}$$

where $V_{\text{eff}}(\mathbf{r})$ is the effective potential which has already been defined, the $|\phi_n\rangle$ are the single-particle wavefunctions with corresponding eigenenergies $\mathcal{E}_n$ and the density is constructed from the single-particle wavefunctions using:

$$n(\mathbf{r}) = \sum_{n=1}^{N} |\phi_n(\mathbf{r})|^2. \tag{2.29}$$

In order to find the total energy, it is therefore necessary to solve for this fictitious non-interacting system using the above set of independent-particle equations, known as the Kohn-Sham equations. In this way, the problem has been replaced by one which is much simpler to solve than the original fully interacting problem.

The Kohn-Sham equations must be solved self-consistently due to the dependence of the effective potential on the density via the Hartree potential. This involves calculating the potential from a given input density, solving the Schrödinger equations to obtain the single-particle orbitals and then using them to construct a new density via Eq. (2.29). This process is repeated until the input and output densities are equal, i.e. the potential and density are self-consistent.

Once the set of Schrödinger equations have been solved in this manner, the ground state energy is then calculated by adding the different energy contributions, with corrections for double-counting. The final expression becomes:

$$E = \sum_{n}^{N} \mathcal{E}_n + E_{\text{xc}}[n] - \int V_{\text{xc}}(\mathbf{r})\, n(\mathbf{r})\, d\mathbf{r} - \frac{1}{2} \iint \frac{n(\mathbf{r})\, n(\mathbf{r}')}{|\mathbf{r} - \mathbf{r}'|}\, d\mathbf{r}\, d\mathbf{r}', \quad (2.30)$$

where the sum over the single-particle eigenvalues is often referred to as the band structure energy.

2.2.3 Exchange and Correlation

Whilst DFT is in principle an exact formalism, as previously mentioned, the exchange correlation energy term must in practice be approximated. The first suggestion for an approximate exchange correlation functional was the local density approximation (LDA), originally proposed by Kohn and Sham [8]. For the rest of this work the LDA has been used, and so this will be described in the following section. A number of other exchange-correlation functionals have since been proposed, which will be only briefly summarized.

2.2.3.1 The Local-Density Approximation

The LDA is aimed at systems where the density is slowly varying and therefore assumes that the system is locally homogeneous, despite being inhomogeneous over-all. Using this assumption, an expression for the exchange-correlation energy is derived from that of a uniform electron gas, which has been calculated to a high level of accuracy using Monte Carlo methods [13] and perturbation theory. The LDA is therefore defined as:

$$E_{\text{xc}}^{\text{LDA}}[n] = \int \epsilon_{\text{xc}}(n(\mathbf{r}))\, n(\mathbf{r})\, d\mathbf{r}, \quad (2.31)$$

where $\epsilon_{\text{xc}}(n(\mathbf{r}))$ is the exchange-correlation energy per particle of a uniform electron gas with density $n(\mathbf{r})$.

At first glance this is a rather crude approximation, and so might not be expected to yield very accurate results for the large number of systems where the density does not vary slowly. However, in practice it has been seen to work remarkably well for a wide range of systems. This surprising success can be partly explained by the fact that the LDA obeys the sum rule for the exchange-correlation hole, more information on which can be found in e.g. [1, 14].

2.2.3.2 Other Functionals

Another popular class of exchange-correlation functionals are generalized-gradient approximations (GGAs), such as [15–17], which aim to take into account the gradient of the density at a given point, as well as the density itself. In many cases GGAs give increased levels of accuracy compared to the LDA, however improvement is not shown in all cases.

A number of other types of exchange-correlation functionals also exist, including so-called hybrid functionals, which combine an orbital-dependent Hartree-Fock term with a purely density based term. One of the most popular examples is B3LYP [18], which is particularly widely used by chemists.

There is no perfect functional, but some functionals are particularly suited to calculating different properties. For example GGAs are expected to get better binding energies and bond lengths than the LDA and B3LYP gives excellent results for molecules, whereas the LDA is still generally slightly better at describing semiconductors. For a summary of some of the different functionals see [1, 2].

2.2.4 The Meaning of the Kohn–Sham Eigenvalues

There are, however, some limitations of DFT which one should be aware of when using it to calculate properties of real systems, particularly when close agreement with experiment is required. In this work we are interested in calculating conduction states for use in the generation of spectra, therefore of particular relevance is the relation between the Kohn-Sham states and the true quasiparticle energies.[1] As can be seen from the derivation of the Kohn-Sham equations, it is clear that there is no obvious correspondence between the two. This has particular implications for the calculation of band gaps using DFT, which are chronically underestimated, and indeed would not be expected to be correct even if one had access to the "true" exchange-correlation functional.

Rather than calculating the Kohn-Sham band gap, i.e. the difference between the lowest unoccupied molecular orbital (LUMO) energy and the highest occupied molecular orbital (HOMO) energy, one should instead use the relation between the gap and the chemical potentials of both the N and $N + 1$ electron systems, $\mu^{(N)}$ and $\mu^{(N+1)}$ respectively [19–21], whereby:

$$E_{\text{gap}} = \mu^{(N+1)} - \mu^{(N)}. \tag{2.32}$$

One can also show that the energy of the HOMO has meaning, by relating it to the chemical potential [22]:

$$\mu^{(N)} = \frac{\partial E^{(N)}}{\partial N} = \varepsilon_N^{(N)}, \tag{2.33}$$

---

[1] Electrons in a crystal do not behave in the same way as free electrons, therefore we treat these effects by working in terms of weakly interacting quasiparticles, rather than the strongly interacting electrons.

where $E^{(N)}$ is the total energy of the N electron system and $\mathcal{E}_N^{(N)}$ is the HOMO energy. Combining Eqs. (2.32) and (2.33), the band gap can therefore be calculated as a difference in HOMO energy levels for the N and $N + 1$ electron systems:

$$E_{\text{gap}} = \mu^{(N+1)} - \mu^{(N)} = \mathcal{E}_{N+1}^{(N+1)} - \mathcal{E}_N^{(N)}, \qquad (2.34)$$

where $\mathcal{E}_{N+1}^{(N+1)}$ is the HOMO for the $N + 1$ electron system. For further discussion on the discrepancy between the DFT band gap and the true band gap, see e.g. [23, 24].

From the above results one would not expect to be able to use the Kohn-Sham states for the calculation of experimental spectra, and indeed their use cannot be rigorously justified. However, in practice reasonable agreement has been seen with experiment, particularly when the scissor operator approximation [25, 26] is employed. Furthermore, as the emphasis within this work is on application to large systems, more accurate methods such as the GW method [27–29] are prohibitively expensive, and so the approximation becomes justified with respect to the aims of studying previously inaccessible system sizes whilst maintaining a reasonable standard of accuracy.

2.3 Periodic Systems

In practice, when solving for periodic interacting systems using DFT or equivalent methods, rather than applying open boundary conditions to systems with a very large number of atoms, it is often more convenient to instead consider a smaller repeat unit and apply periodic boundary conditions. This has the obvious benefit of reducing an essentially infinite system to a finite one, and so for the rest of this work, we will be working with such boundary conditions.

2.3.1 Supercells

The benefits of applying periodic boundary conditions to solids are quite clear, particularly for regular solids where the unit cell requires only a few atoms. However, one can also apply periodic boundary conditions to solids with point defects, for example, or to molecules. This is achieved via the use of the supercell approximation, which for a molecular system involves placing the molecule in a large unit cell with sufficient empty space surrounding it so that the molecule does not interact with any of its periodic images. For a defect system, for example, this would involve placing the unit cell containing the defect in the centre of a much larger cell, and surrounding it with unchanged unit cells such that the defect does not interact with any of its periodic copies. Both of these examples are depicted in Fig. 2.1.

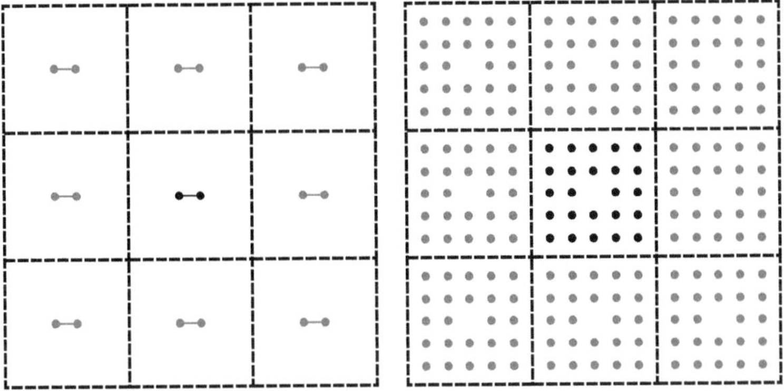

Fig. 2.1 Schematics demonstrating the use of the supercell approximation for a molecule (*left*) and a vacancy defect (*right*). The *dotted lines* indicate the cell boundaries, and the atoms shown in *grey* are periodic images of the cell

2.3.2 Bloch's Theorem and k-Point Sampling

We consider here the case of an electron in a perfect crystal, with an infinitely repeating potential defined by the Bravais lattice of the nuclei. If we take a general Bravais lattice vector $\mathbf{R}$, it can be written in terms of the primitive lattice vectors $\{\mathbf{a}_i\}$, by:

$$\mathbf{R} = n_1\mathbf{a}_1 + n_2\mathbf{a}_2 + n_3\mathbf{a}_3. \tag{2.35}$$

The potential therefore satisfies:

$$V\left(\mathbf{r} + \mathbf{R}\right) = V\left(\mathbf{r}\right). \tag{2.36}$$

Restricting ourselves to the case of non-interacting electrons, which corresponds to the fictional non-interacting Kohn-Sham system in the case of DFT, we shall now derive Bloch's theorem, which has important consequences for calculations on periodic systems. The appropriate Schrödinger equation will be:

$$\hat{H}|\psi\rangle = \left(-\frac{1}{2}\nabla^2 + V\left(\mathbf{r}\right)\right)|\psi\rangle = \mathcal{E}|\psi\rangle, \tag{2.37}$$

where $\hat{H}$ will retain the periodicity of the crystal. We next define a translation operator,[2] $\hat{T}_\mathbf{R}$, which acts on any function of position as follows:

$$\hat{T}_\mathbf{R} f\left(\mathbf{r}\right) = f\left(\mathbf{r} + \mathbf{R}\right). \tag{2.38}$$

---

[2] We note that the translation operator is not a Hermitian operator and it does not correspond to any observables of the system.

As the Hamiltonian follows the same periodicity, it will commute with the translation operator. Furthermore, the act of applying two successive translation operators corresponds to another translation, so that the order of operation is irrelevant. This can be expressed mathematically by the following relations:

$$\hat{T}_{\mathbf{R}}\hat{T}_{\mathbf{R}'} = \hat{T}_{\mathbf{R}'}\hat{T}_{\mathbf{R}} = \hat{T}_{\mathbf{R}+\mathbf{R}'}. \tag{2.39}$$

Based on the fact that the Hamiltonian and translation operator commute, one can write down a set of states which are eigenstates of both operators, so that:

$$\hat{H}|\psi\rangle = \mathcal{E}|\psi\rangle \tag{2.40}$$

and

$$\hat{T}_{\mathbf{R}}|\psi\rangle = c\,(\mathbf{R})\,|\psi\rangle. \tag{2.41}$$

The eigenvalues can be rewritten in terms of the primitive lattice vectors and three numbers $\{x_i\}$:

$$c\,(\mathbf{a}_i) = e^{2\pi i x_i}, \quad i = 1, 2, 3. \tag{2.42}$$

The $\{x_i\}$ are in general complex. However, under the application of periodic boundary conditions, it can be shown that they must be real, which allows one to define the set of allowed $\mathbf{k}$-vectors. Using Eq. (2.39), the eigenvalues of the translation operator can be related via:

$$\hat{T}_{\mathbf{R}}\hat{T}_{\mathbf{R}'}|\psi\rangle = c\,(\mathbf{R})\,c\,\left(\mathbf{R}'\right)|\psi\rangle = c\,\left(\mathbf{R}+\mathbf{R}'\right)|\psi\rangle, \tag{2.43}$$

which agrees with the properties of exponential functions. Given the definition of $\mathbf{R}$ in Eq. (2.35) and the eigenvalue relations in Eq. (2.43), we can therefore write:

$$c\,(\mathbf{R}) = c\,(\mathbf{a}_1)^{n_1}\,c\,(\mathbf{a}_2)^{n_2}\,c\,(\mathbf{a}_3)^{n_3}\,. \tag{2.44}$$

If we now consider a vector $\mathbf{k}$ in reciprocal space, which we define in terms of the reciprocal lattice vectors $\{\mathbf{b}_i\}$:

$$\mathbf{k} = x_1\mathbf{b}_1 + x_2\mathbf{b}_2 + x_3\mathbf{b}_3, \tag{2.45}$$

where the reciprocal lattive vectors follow the relation $\mathbf{b}_i\cdot\mathbf{a}_j = 2\pi\delta_{ij}$, we can rewrite Eq. (2.44) as:

$$c\,(\mathbf{R}) = e^{i\mathbf{k}\cdot\mathbf{R}}. \tag{2.46}$$

Returning to the eigenvalue Eq. (2.41) and combining with the definition of the translation operator in Eq. (2.38), we now have:

$$\hat{T}_{\mathbf{R}}\psi\,(\mathbf{r}) = \psi\,(\mathbf{r}+\mathbf{R}) = c\,(\mathbf{R})\,\psi = e^{i\mathbf{k}\cdot\mathbf{R}}\psi\,(\mathbf{r})\,. \tag{2.47}$$

At this point it is logical to label the wavefunction with the appropriate wavevector $\mathbf{k}$ and band index n, which refers to the independent eigenstates which will occur for each value of $\mathbf{k}$. We therefore now have one form of Bloch's theorem:

$$\psi_{n\mathbf{k}}\left(\mathbf{r}+\mathbf{R}\right) = e^{i\mathbf{k}\cdot\mathbf{R}}\psi_{n\mathbf{k}}\left(\mathbf{r}\right). \tag{2.48}$$

This can be written in an alternative form by considering the function $u\left(\mathbf{r}\right)$, defined as:

$$u_{n\mathbf{k}}\left(\mathbf{r}\right) = e^{-i\mathbf{k}\cdot\mathbf{r}}\psi_{n\mathbf{k}}\left(\mathbf{r}\right), \tag{2.49}$$

which can be shown to have the periodicity of the lattice using Eq. (2.48). Rearranging Eq. (2.49) we have:

$$\psi_{n\mathbf{k}}\left(\mathbf{r}\right) = e^{i\mathbf{k}\cdot\mathbf{r}}u_{n\mathbf{k}}\left(\mathbf{r}\right), \tag{2.50}$$

which is the second version of Bloch's theorem. See Ashcroft and Mermin [30] for an alternative proof of the theorem.

We can now use Bloch's theorem to show that any wave vector can be related to another wave vector inside the first Brillouin zone. We first write such a general vector $\mathbf{k}'$ as:

$$\mathbf{k}' = \mathbf{k} + \mathbf{G}, \tag{2.51}$$

where $\mathbf{G}$ is some linear combination of the reciprocal lattice vectors $\{\mathbf{b}_i\}$ and $\mathbf{k}$ lies in the first Brillouin zone. We then substitute Eq. (2.51) into the second form of Bloch's theorem (Eq. 2.50), giving:

$$\begin{aligned}
\psi_{n\mathbf{k}'}\left(\mathbf{r}\right) &= e^{i\mathbf{k}'\cdot\mathbf{r}}u_{n\mathbf{k}'}\left(\mathbf{r}\right) = e^{i\mathbf{k}\cdot\mathbf{r}}\left[e^{i\mathbf{G}\cdot\mathbf{r}}u_{n\mathbf{k}'}\left(\mathbf{r}\right)\right] \\
&= e^{i\mathbf{k}\cdot\mathbf{r}}u_{n'\mathbf{k}}\left(\mathbf{r}\right) = \psi_{n'\mathbf{k}}\left(\mathbf{r}\right).
\end{aligned} \tag{2.52}$$

We have now transformed the problem from one of obtaining an infinite number of wavefunctions to one of obtaining the finite set of occupied wavefunctions at those $\mathbf{k}$-points which lie within the first Brillouin zone. Of course, for the limit of an infinite periodic system, there will still be an infinite number of $\mathbf{k}$-points lying within the first Brillouin zone. However, the wavefunctions and eigenvalues of the system are known to vary smoothly with respect to $\mathbf{k}$, therefore it is possible to consider only a finite set of $\mathbf{k}$-points. There are a number of methods of effectively sampling the Brillouin zone in this manner, e.g. those discussed in [31, 32] and one can also take advantage of symmetry relations. Additionally, one could use $\mathbf{k}\cdot\mathbf{p}$ perturbation theory [33–35], which will be returned to in Sect. 5.2.1, to approximate the solutions at neighbouring $\mathbf{k}$-points to those already calculated and therefore integrate over the Brillouin zone more accurately. Finally we note that for large supercells the volume of the Brillouin zone, Ω_{BZ}, becomes very small, as its volume is defined by:

$$\Omega_{BZ} = \frac{(2\pi)^3}{\Omega_{cell}},\qquad(2.53)$$

where Ω_{cell} is the volume of the supercell. Thus only a small number of **k**-points need to be included to accurately sample the Brillouin zone. For the purposes of this work, where the goal is to study large systems, we will restrict ourselves to sampling the Γ-point only, which also allows us to require the wavefunctions to be real. For further discussions on Bloch's theorem and its consequences see for example [2, 30, 36].

2.3.3 Plane-Waves

In order to solve the appropriate eigenvalue equation for a periodic system, in our case the Kohn-Sham equations, we require the use of a basis set to represent the wavefunction. We will return to the issue of basis sets in Sect. 3.4 and Chap. 5, in particular localized basis sets. However, one of the most commonly used basis sets for calculations on extended systems is the plane-wave basis set, which is a natural choice following on from the second version of Bloch's theorem (Eq. 2.50). The periodic function $u_{nk}(\mathbf{r})$ is Fourier expanded so that the wavefunction becomes a sum over plane-waves:

$$\psi_{nk}(\mathbf{r}) = e^{i\mathbf{k}\cdot\mathbf{r}} \sum_{\mathbf{G}} c_{n,\mathbf{k}+\mathbf{G}} e^{i\mathbf{G}\cdot\mathbf{r}},\qquad(2.54)$$

where the $c_{n,\mathbf{k}+\mathbf{G}}$ are the expansion coefficients. In principle, this is an infinite sum over the reciprocal lattice vectors, however in practice it is possible to truncate the sum so that only those components up to some kinetic energy cut-off, E_{cut}, are included. This truncation is defined by:

$$E_{cut} \leq \frac{|\mathbf{k}+\mathbf{G}|^2}{2}.\qquad(2.55)$$

In order to successfully employ such a truncation, one must systematically converge the final results with respect to the cut-off energy; this convergence of total energy is variational with increasing cut-off energy, which is a particular advantage of the plane-wave basis set.

There are a number of other advantages of plane-waves, including the fact that they are orthogonal and highly efficient algorithms exist for their manipulation, in particular fast Fourier transforms (FFTs), which are used to convert between real and reciprocal space. In addition, they are unbiased and cover all space evenly, which is advantageous for the calculation of forces. However, this is also a disadvantage in that empty space has a high cost associated with it. Furthermore, a very large number of basis functions are required for accurate results compared with other basis sets. Finally, plane-waves are not a natural choice for the representation of quickly varying functions, which means that the cost of representing the wavefunction near the ionic cores is prohibitively high. In order to overcome this obstacle, it is customary to employ the pseudopotential approximation, as described in the following section.

2.3.4 Pseudopotentials

The pseudopotential approximation aims to overcome the problems with representing the wavefunction near the core of an atom, where they are highly oscillatory and thus require a large number of plane-waves to accurately describe them. Rather than using atomic potentials and therefore trying to represent the "true" wavefunction, one instead uses an effective potential, or *pseudopotential*, and therefore need only be able to represent a *pseudowavefunction* [37, 38]. This has the advantage that much weaker potentials than the strong Coulomb potential can be used, resulting in smoother and less oscillatory pseudowavefunctions. This is achieved by absorbing the effect of the core states into the pseudopotential, only representing the valence states explicitly. This is justified via the observation that the core states are only weakly affected by their local chemical environment, and so one can expect that they would remain fairly similar whether in an isolated atom or in an extended crystal.

There are a number of different types of pseudopotentials and a variety of methods for generating them, which can broadly be assigned into two categories: norm-conserving [39], which retain the correct scattering properties to first order in energy, and ultrasoft [40]. Norm-conserving pseudopotentials have stricter requirements in their generation than ultrasoft pseudopotentials, thus ultrasofts are smoother and so require lower cut-off energies than norm-conserving pseudopotentials, although the complexity of the expressions involved becomes more complicated. A schematic for an example pseudopotential is shown in Fig. 2.2.

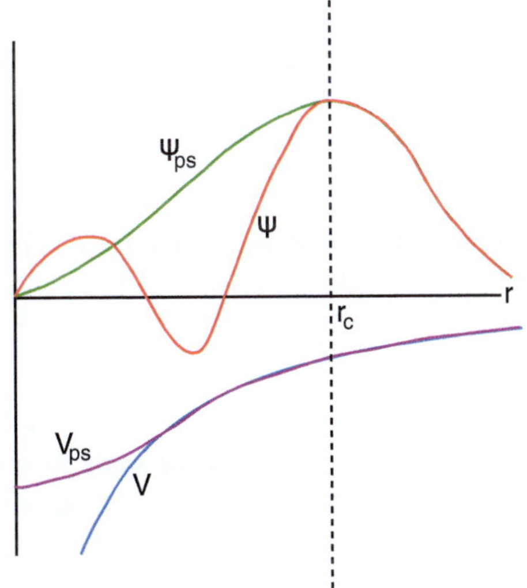

Fig. 2.2 Diagram depicting the pseudopotential approximation; example pseudopotentials and pseudowavefunctions are plotted with the original wavefunction and potential, where the pseudo quantities agree with the original quantities outside some cut-off radius r_c

In this work we will be using norm-conserving pseudopotentials, in the separable form of Kleinman and Bylander [41]. This type of pseudopotential is fully non-local and its action is both energy and angular momentum dependent. The pseudopotential operator $\hat{V}_{\mathrm{KB}}$ is of the form:

$$\hat{V}_{\mathrm{KB}} = \hat{V}_{\mathrm{loc}} + \sum_{l} \sum_{m=-l}^{l} \frac{|\delta\hat{V}_l \chi_{lm}\rangle\langle\chi_{lm}\delta\hat{V}_l|}{\langle\chi_{lm}|\delta\hat{V}_l|\chi_{lm}\rangle}, \qquad (2.56)$$

where $\{|\chi_{lm}\rangle\}$ are the pseudo-orbitals, $\hat{V}_{\mathrm{loc}}$ is the local part of the atomic pseudopotential and $\delta\hat{V}_l = \hat{V}_l - \hat{V}_{\mathrm{loc}}$, with $\hat{V}_l$ being the angular momentum dependent part of the pseudopotential. This is constructed so that:

$$\hat{V}_{\mathrm{KB}}|\chi_{lm}\rangle = \hat{V}_l|\chi_{lm}\rangle, \qquad (2.57)$$

i.e. its action on the reference pseudowavefunctions is the same as an equivalent semi-local pseudopotential. One of the chief advantages of this form is that the number of operations required to calculate the pseudopotential matrix elements can be made to scale linearly with the number of plane-waves, whereas for semi-local forms this calculation scales quadratically. This combination of a plane-wave basis set with the use of pseudopotentials is referred to as the plane-wave pseudopotential method (PWPP), and it will be used as the standard of comparison within this work.

2.4 Summary

DFT has become a very popular method for electronic structure calculations, having the advantage over tight-binding methods that it is *ab initio*, whilst being more efficient than other methods such as Hartree-Fock and Monte Carlo methods. However, there is no systematic method for improving the exchange-correlation functional, which can be a problem for some systems where current functionals are not sufficiently accurate. Despite this, it has proven to be a hugely successful method for a large number of systems.

The development of computationally efficient codes based on the PWPP method in particular has allowed for a large number of calculations determining a wide variety of electronic properties for an extensive list of systems. However, in this form DFT scales cubically with the number of atoms, for reasons which we shall explain in the following chapter, and so the system sizes which can be treated are inherently limited. In the following chapter, we discuss the development of various methods which improve upon this standard cubic-scaling behaviour. These linear-scaling methods allow for the study of systems of very large numbers of atoms, particularly with increasing computing capabilities, and so are of vital importance in extending the scope of DFT.

References

1. Martin, R.: Electronic Structure: Basic Theory and Practical Methods. Cambridge University Press, Cambridge (2004)
2. Kohanoff, J.: Electronic Structure Calculations for Solids and Molecules: Theory and Computational Methods. Cambridge University Press, Cambridge (2006)
3. Wimmer, E.: Mater. Sci. Eng. B **37**(1–3), 72 (1996)
4. Born, M., Oppenheimer, R.: Ann. Phys. **389**(20), 457 (1927)
5. Thomas, L.H.: Math. Proc. Cambridge Philos. Soc. **23**(05), 542 (1927)
6. Fermi, E.: Z. Phys. **48**(1), 73 (1928)
7. Hohenberg, P., Kohn, W.: Phys. Rev. **136**(3B), B864 (1964)
8. Kohn, W., Sham, L.J.: Phys. Rev. **140**(4A), A1133 (1965)
9. Levy, M.: Phys. Rev. A **26**(3), 1200 (1982)
10. Levy, M.: Proc. Natl. Acad. Sci. U.S.A. **76**(12), 6062 (1979)
11. Lieb, E.: In: Shimony, A., Feshbach, H. (eds.) Physics as Natural Philosophy: Essays in Honor of Laszlo Tisza on His 75th Birthday, pp. 111–149. MIT Press, Cambridge (1982)
12. Lieb, E.H.: Int. J. Quantum Chem. **24**(3), 243 (1983)
13. Ceperley, D.M., Alder, B.J.: Phys. Rev. Lett. **45**(7), 566 (1980)
14. Perdew, J.P., Zunger, A.: Phys. Rev. B **23**, 5048 (1981)
15. Perdew, J.P., Burke, K., Ernzerhof, M.: Phys. Rev. Lett. **77**, 3865 (1996)
16. Becke, A.D.: Phys. Rev. A **38**, 3098 (1988)
17. Perdew, J.P., Wang, Y.: Phys. Rev. B **45**, 13244 (1992)
18. Becke, A.D.: J Chem. Phys. **98**(7), 5648 (1993)
19. Perdew, J.P., Levy, M.: Phys. Rev. Lett. **51**, 1884 (1983)
20. Sham, L.J., Schlüter, M.: Phys. Rev. Lett. **51**, 1888 (1983)
21. Sham, L.J., Schlüter, M.: Phys. Rev. B **32**, 3883 (1985)
22. Janak, J.F.: Phys. Rev. B **18**, 7165 (1978)
23. Godby, R.W., Schlüter, M., Sham, L.J.: Phys. Rev. B **37**, 10159 (1988)
24. Jones, R.O., Gunnarsson, O.: Rev. Mod. Phys. **61**, 689 (1989)
25. Godby, R.W., Schlüter, M., Sham, L.J.: Phys. Rev. Lett. **56**(22), 2415 (1986)
26. Gygi, F., Baldereschi, A.: Phys. Rev. Lett. **62**(18), 2160 (1989)
27. Hedin, L.: Phys. Rev. **139**(3A), A796 (1965)
28. Godby, R.W., Needs, R.J.: Phys. Scr. **T31**, 227 (1990)
29. Del Sole, R., Reining, L., Godby, R.W.: Phys. Rev. B **49**(12), 8024 (1994)
30. Ashcroft, N.W., Mermin, N.D.: Solid State Physics. Holt, Rinehart and Winston, New York (1976)
31. Monkhorst, H.J., Pack, J.D.: Phys. Rev. B **13**, 5188 (1976)
32. Chadi, D.J., Cohen, M.L.: Phys. Rev. B **8**, 5747 (1973)
33. Robertson, I.J., Payne, M.C.: J. Phys.: Condens. Matter **2**(49), 9837 (1990)
34. Robertson, I.J., Payne, M.C.: J. Phys.: Condens. Matter **3**(45), 8841 (1991)
35. Pickard, C.J., Payne, M.C.: Phys. Rev. B **62**(7), 4383 (2000)
36. Kittel, C.: Introduction to Solid State Physics, 7th edn. John Wiley and Sons, New York (1996)
37. Phillips, J.C., Kleinman, L.: Phys. Rev. **116**, 287 (1959)
38. Phillips, J.C.: Phys. Rev. **112**, 685 (1958)
39. Hamann, D.R., Schlüter, M., Chiang, C.: Phys. Rev. Lett. **43**, 1494 (1979)
40. Vanderbilt, D.: Phys. Rev. B **41**, 7892 (1990)
41. Kleinman, L., Bylander, D.M.: Phys. Rev. Lett. **48**, 1425 (1982)

Chapter 3
Linear-Scaling Methods

As described in the previous chapter, DFT is a very popular method for electronic structure calculations which uses the electronic charge density as the central variable, thereby reducing the number of variables required to solve the problem. This allows calculations on interacting systems of electrons which would not otherwise be possible. However, the maximum system size that can be studied is still limited by the cubic or $\mathcal{O}\left(\mathcal{N}^3\right)$ scaling of the standard formulation of Kohn-Sham DFT, where $\mathcal{N}$ is the number of atoms in the system. This $\mathcal{O}\left(\mathcal{N}^3\right)$ scaling does not reflect the locality of the system as implied by the quantum-mechanical principle of "near-sightedness". In particular, the density matrix (DM) is known to be exponentially localized in non-metallic systems. This knowledge has led to the development of a number of methods which take advantage of this localization and thus scale linearly with system size. Such methods are known as order $\mathcal{N}$, or $\mathcal{O}\left(\mathcal{N}\right)$.

In this chapter, we will first briefly consider some of the motivations for and possible applications of linear-scaling methods. We will then outline the origins of the $\mathcal{O}\left(\mathcal{N}^3\right)$ scaling and the justification for linear-scaling methods, followed by a summary of a number of different methods of linear-scaling DFT. Finally we will present the methodology used in the ONETEP [1–3] code, which is the code used within this work.

3.1 Motivation and Applications

The standard formulation of DFT can be made to be highly efficient for small systems, however the asymptotic cubic scaling becomes a problem for larger systems, limiting the sizes of systems that can be studied within current computational capabilities. This limit is currently set at system sizes approaching a thousand atoms. For linear-scaling methodologies, it is important to consider the prefactor of the scaling, as this is almost always much higher than standard DFT calculations. This leads to the concept of a crossover point, which is the system size for which linear-scaling methods become more efficient than standard methods. It would be pointless for

L. Ratcliff, *Optical Absorption Spectra Calculated Using Linear-Scaling Density-Functional Theory*, Springer Theses, DOI: 10.1007/978-3-319-00339-9_3,
© Springer International Publishing Switzerland 2013

example to have a methodology which has such a high prefactor that the crossover point is higher than can actually be accessed within current computational limits! Thus it is always necessary to consider where the crossover point is likely to be for a given method. Although this crossover point will depend on both the approach taken and the system being studied (in particular the dimensionality of the system), this crossover typically occurs at system sizes of around 1000 atoms, thus linear-scaling methods become highly applicable for systems around this size and higher.

Linear-scaling methods are designed to take full advantage of parallel supercomputers. This has allowed benchmark calculations of for example over 30,000 atoms of bulk Si [3], more than one million atoms of bulk aluminium [4] and more than 11 million atoms of alumina [5], none of which would be accessible to conventional DFT formulations.

This ability to study larger systems is particularly relevant for large biomolecules or nanosystems. Examples of systems which have already been studied by various linear-scaling methods include, Si nanowires [6], bulk Si [7, 8], biomolecules [9, 10], CdSe nanorods [5] and fullerenes [11]. More generally, any calculation which requires the use of a large unit cell or uses the supercell method will greatly benefit from $\mathcal{O}(\mathcal{N})$ methods.

3.1.1 Cubic-Scaling Behaviour and Nearsightedness

One of the most common implementations of DFT uses a plane-wave basis and pseudopotentials, as discussed in the previous chapter. Much work has been done on improving the computational efficiency of such methods, however in this form and in all other standard formulations, DFT has been shown to scale asymptotically as the cube of the number of atoms. For small systems, this is not a problem but this does impose a strict limit on the number of atoms that can be treated due to current computational capabilities. The origins of this cubic scaling are quite easy to see; the number of basis functions will scale as $\mathcal{O}(\mathcal{N})$ with the number of atoms, however the need to impose orthogonality constraints on the solutions will become the dominating factor at large $\mathcal{N}$. This is because the number of wavefunctions increases as $\mathcal{O}(\mathcal{N})$; each wavefunction must be orthogonalized to all others, requiring a number of scalar products which scales as $\mathcal{O}(\mathcal{N})$; and finally the effort required to perform each scalar product will scale as $O(\mathcal{N})$ due to the increasing number of basis functions. Therefore the three $\mathcal{O}(\mathcal{N})$ terms will combine to give $\mathcal{O}(\mathcal{N}^3)$.

The principle of nearsightedness [12, 13] is a well established quantum mechanical principle which explains why this $\mathcal{O}(\mathcal{N}^3)$ scaling need not exist, that in fact linear-scaling DFT ought to be possible. It states that for quantum mechanical systems containing a large number of interacting particles, physical processes are usually only affected by their immediate locality. This means that the information required to solve the Schrödinger equation ought to scale linearly, allowing for linear-scaling DFT formulations. Kohn [12] states that it is generally a consequence of wave-mechanical destructive interference, thus requiring the presence of many particles, although they need not all be interacting. It is generally but not universally valid, for example it does not hold in systems of non-interacting three dimensional bosons below the critical point.

For both insulating and metallic systems, this locality has been quantified for the density matrix, ρ, which is known to decay exponentially with respect to distance for systems with a band gap [14–16]:

$$\rho\left(\mathbf{r}, \mathbf{r}'\right) \sim e^{-\gamma|\mathbf{r}-\mathbf{r}'|}, \tag{3.1}$$

where γ is positive. In metallic systems this decay is less strong, following a power law rather than exponential decay. One therefore ought to be able to take advantage of this principle in order to develop linear-scaling formalisms of DFT, and indeed a variety of such methods exist, which we shall summarize in the following section.

3.2 Overview of Linear-Scaling Methods

It has been known for some time that the $\mathcal{O}\left(\mathcal{N}^3\right)$ scaling of the standard implementation of DFT is a consequence of the methodologies used, rather than the inherent physics of the system. As a result, the idea of forming a linear-scaling formulation of DFT is also not a new one, and so linear-scaling formalisms for both the tight-binding and DFT methods have existed for well over a decade. Consequently, there are a large variety of different methods, concerning which there have been a number of reviews [17–22].

Linear-scaling methods were formed into six main categories by Goedecker [17]: the Fermi operator expansion, the Fermi operator projection, the divide and conquer method, the density matrix minimization method, the orbital minimization method and the optimal basis density matrix minimization method. Alternatively, Stephan and Drabold [23] identified three main categories; variational and differential methods for the calculation of the DM, variational calculation of Wannier functions and non-variational calculation involving the use of projectors. Artacho et al. [24] applied the categories of quantum chemistry methods which use Gaussian basis sets and apply thresholds, physics methods which vary more in their methodologies and hybrid methods which use atomic orbitals with plane-waves as an auxiliary basis. However, there are also some methods which do not fall neatly into any of these schemes, therefore we shall not be applying the same categories.

One fairly obvious division between methods is those which can be applied to metals, and those which are applicable to semi-conductors and insulators only. This is due to the fact that the decay of the DM is only exponential for non-metals, instead obeying a power law for metals, as has been stated in Sect. 3.1.1. This exponential decay is fundamental to most linear-scaling approaches, including ONETEP, thus limiting their application to non-metals. We will therefore first discuss those methods which can be applied to metals, before moving on to the bulk of methods, which are restricted to non-metals. There can also be cases in which certain methodologies are more suited to specific systems, for example large molecules or random alloys.

Before discussing some of the most common methods, we shall first highlight some important factors to consider when discussing linear-scaling methods. Firstly, some so-called linear-scaling methods do not asymptotically scale linearly, for example some of the earlier methods scale as $\mathcal{O}\left(\alpha\mathcal{N} + \beta\mathcal{N}^2\right)$ and therefore asymptotically scale quadratically. As $\alpha \gg \beta$ the limit may not be reached in practice, however as increasingly large systems are studied this could become a limiting factor. Similarly there are also some cases when the asymptotic scaling is strictly $\mathcal{O}\left(\mathcal{N}\log\left(\mathcal{N}\right)\right)$, for example those methods which use FFTs to solve the Poisson equation. In practice for many methods the scaling will be linear for calculations on smaller systems, where those parts of the calculation which are not linear-scaling are insignificant, however for larger calculations these aspects can become dominant. Furthermore, it is also possible for a method to be linear-scaling in principle, in that the individual steps involved scale linearly; but not linear-scaling in practice, due to an increased number of iterations being needed for larger systems.

Finally, it is worth mentioning that many linear-scaling DFT methods originated as tight-binding formulations, some of which are discussed in a review by Bowler et al. [25], which also includes the application to a range of test systems, such as Si and benzene. The link between DFT and tight-binding is demonstrated by Horsfield and Bratkovsky [26] in their review of some common *ab initio* tight-binding methods. This link means that in many cases the methods can be applied to both DFT and tight-binding, thus we will also be discussing some of these tight-binding methods.

3.2.1 Methods Which can be Applied to Metals

Orbital-Free Density-Functional Theory

One of the most common linear-scaling formalisms of DFT that can be applied to metals is orbital-free DFT (OFDFT). Avoiding the use of orbitals seems a natural way to impose linear-scaling, as the imposition of orthogonality constraints on the orbitals is one of the factors which dominates the $\mathcal{O}\left(\mathcal{N}^3\right)$ scaling of standard DFT. The orbitals only exist in DFT to allow the calculation of the kinetic energy, which is achieved by representing the non-interacting electrons of the fictional system. Therefore the creation of orbital-free kinetic energy functionals eliminates the need for orbitals. The earliest such functional was that created by Thomas and Fermi (see Sect. 2.2), but it is severely limited in its accuracy, applicable only to some plasmas. One slightly more developed functional was that of von Weizsäcker [27] which included a gradient-based term.

Since the development of these early kinetic energy functionals, many more have been created that have been shown to give reasonably accurate results for many metals, although as yet none are capable of correctly modelling either transition metals or non-metals. One example is that of Wang and Teter [28] which is based on an integral term which includes higher order corrections than the von Weizsäcker functional. García-Aldea and Alvarellos [29] have reviewed a selection of semi-local kinetic energy functionals, however one drawback of OFDFT is that there is no provision for fully non-local potentials, as orbitals are required for their implementation. Nonetheless, there are still a number of systems which can be studied without non-local potentials.

OFDFT was used by Pearson et al. [30] with the Perrot functional to model simple metals such as sodium, for which good results were achieved, however for large calculations using this method the $\mathcal{O}\left(\mathcal{N}^2\right)$ scaling of the calculation of the ionic structure factor starts to dominate and thus it is no longer truly linear-scaling.

A more recent formulation of OFDFT has been tested on a system of over one million atoms of bulk aluminium, taking less than 6 h using 192 processors [4]. A new formulation was used to avoid the $\mathcal{O}\left(\mathcal{N}^2\right)$ bottleneck of the structure factor calculation, however it was still only quasi-linear-scaling due to the use of FFTs, which strictly scale as $\mathcal{O}\left(\mathcal{N}\log\left(\mathcal{N}\right)\right)$, rather than $\mathcal{O}\left(\mathcal{N}\right)$. This has been further developed and applied to a number of semiconductors, reproducing properties such as bulk moduli to a reasonable level of accuracy [31].

Multiple Scattering Approach

Wang et al. [32] used a real space multiple scattering theory approach that was used as an all-electron method and applied a muffin-tin orbital or atomic sphere approximation, which restricted its application to systems of metallic alloys.

Abrikosov et al. [33] used electronic multiple scattering processes within a Green's function approach whereby the Green's function matrix is inverted. This is a general method applied within finite regions of space labelled local interaction zones (LIZ) and is applied sequentially to each atom in the unit cell. The size of the LIZ can be reduced by embedding in an effective medium. This is known as a locally self-consistent Green's function approach and is again particularly applicable to random alloys. Another scattering approach is the Korringa Kohn-Rostocker (KKR) method, which is discussed in the context of linear-scaling by Zeller [34]. This again involves a Green's function formulation and uses an exact screening transformation to reduce the computational complexity of the original KKR method. It was applied within tight-binding to calculate total energies for systems of up to 500 atoms of three different metals. More recently this has been applied to metallic supercells of over 30,000 atoms using an iterative method which takes advantage of the sparsity of the underlying matrices [35].

The Divide and Conquer Method

The earliest linear-scaling DFT method was the divide and conquer method by Yang [36], which involved dividing a given system into several smaller subsystems then recombining to give the final result. The subsystems can be formed in any combination, with atoms being allowed to belong to more than one system. The division is done in real-space using a smooth partition function, then a local approximation is made to the Hamiltonian, which is allowed to be different in each subsystem. A localized basis is used and diagonalization is done independently in each system, which also makes parallelization straightforward. The method involves a minimal coupling between systems, through the local potential and value for the Fermi energy only. The sizes of the subsystems are independent of the total system size, which means that the computational effort for a given subsystem is also independent of system size. The divide and conquer method is however non-variational, although it gives the correct Kohn-Sham energy in the limit of a complete basis, and there is also no method to calculate forces consistently, which can lead to small inconsistencies in the energy and forces. There is also a high crossover point, compared to other methods. The divide and conquer method has been applied to metallic-systems, as shown for example by Shimojo et al. [5] who applied it to liquid rubidium.

Two related methods are the fragment molecular orbital approach [37, 38] and the molecular tailoring approach [39, 40], which are designed for application to large molecules. Both of these methods are based around the principle of dividing molecules into smaller fragments, calculating either the molecular orbitals or the electrostatic potential for the fragments, then recombining the fragments to find the result for the complete system.

Other Methods

Other linear-scaling methods that can be applied to metals include that of Stechel et al. [41], which uses a combination of localized orbitals for those states in metals which are occupied throughout the Brillouin zone, the number of which scales as $\mathcal{O}(\mathcal{N})$, and delocalized orbitals for those states which cross the Fermi level, the number of which is expected to be independent of system size. This method was expected to be accurate for both metals and insulators, and was also expected to be more efficient than some other methods as only information about occupied states is required, however it was only tested within a tight-binding context, rather than full DFT.

The Fermi operator expansion method (FOE) involves defining the DM as a matrix functional of the Hamiltonian and finding a computable form [17, 42]. This involves calculating the full DM and thus is less efficient than some other methods, however it does scale linearly with the size of the localization region and has a relatively low crossover point as well as quite low memory requirements. In its basic form this method applies to insulators only but Baer and Head-Gordon [43] devised a version that can be applied to metals. This involved a renormalization-group approach, whereby the density operator is written as a telescopic series with a decreasing temperature, allowing the energy of the system to be rescaled by the same factor as the temperature scaling. At finite temperatures the DM is more localized for metals thus this approach avoids the restriction to non-metallic systems only. It was applied within tight-binding to carbon nanotubes and was shown to exhibit quasi-linear-scaling.

Krajewski and Parrinello [44, 45] developed a stochastic method which can be applied to both metallic and insulating systems. It was originally applied to one-dimensional systems, however it has more recently been applied to bulk systems as well [46].

3.2.2 Methods Which are Restricted to Non-Metals

Statistical Approach

One approach to linear-scaling by Drabold and Sankey [47] involves a statistical approach to estimate the entire eigenvalue spectrum. Instead of relying on a local-ization approximation, like most $\mathcal{O}(\mathcal{N})$ methods, a statistical sampling is instead used to obtain sufficient information to accurately construct the electronic density of states (DOS), and thus also the band structure energy. The information in the DOS is contained in a so-called impartial vector, which is generated from random vectors using a penalty functional to tend towards the correct impartial vector. This is then converted into the DOS using the maximum entropy principle to give the best guess by considering it to be a probability density. The individual eigenvalues obtained are not exact, however the computed DOS is correct to a good level of approxima-tion. One problem with the application of random vectors is that it can result in the appearance of random noise, and also the calculation of forces is quite expensive.

Projection Methods

Stephan and Drabold [23] described an $\mathcal{O}(\mathcal{N})$ scheme which takes advantage of the projection properties of the density operator. The main feature is the representation of the projection operator by a polynomial, the degree of which must be higher for smaller band gaps, highlighting the difficulty with applying to metals. The density operator, $\hat{\rho}$, is written as:

$$\hat{\rho} = 2F\left(\hat{H}\right),\tag{3.2}$$

where $F(E) = \left[e^{(E-\mu)/kT} + 1\right]^{-1}$ is the Fermi distribution at temperature T and μ is the chemical potential. An initial estimate is required for both the Fermi energy of the system and band edges, however these can be extracted from other methods such as the recursion method or the maximum entropy scheme, thus retaining the linear-scaling. One particular advantage is that no initial guess is required for either the density matrix or functions; this gives a strong advantage for systems with high disorder within molecular dynamics calculations. Goedecker and Colombo [42, 48] used a similar projection based method, which uses a polynomial approximation to the Fermi distribution between the lowest and highest eigenvalues. They note that ordered metals could also be treated with the method if a finite temperature is applied, however the approach will scale quadratically; extremely large system sizes are required before linear-scaling behaviour will be seen.

Orbital Minimization Methods

Mauri and Galli [49] use an approach where they define an energy functional which avoids the need to impose explicit orthogonalization constraints, therefore also avoiding the need to calculate the inverse overlap matrix, a procedure which is relatively expensive. They also use a basis of localized orbitals, reducing the overall procedure to a succession of calculations of products of sparse matrices, which can be made to be efficient. Shimojo [50] also used this functional in a real-space grid approach involving the use of high order finite difference methods for the calculation of the derivatives necessary for the kinetic energy.

Ordejon et al. [51] employ a similar functional method, where they use Lagrange multipliers to define a functional which imposes orthogonality constraints. The same family of functionals can be derived using a Taylor series expansion [52] and a recursive relation can be used to avoid deviation from the orthogonality constraints. The lack of constraints improves both the robustness and the scaling of the algorithm, although there can be some problems with shallow local minima. This issue can be solved using the generalization of Kim et al. [53], who use a higher number of orbitals than there are electrons, thereby increasing the stability of the algorithm. Whilst the eigenvalues are not explicitly found, they can be calculated afterwards.

SIESTA [24, 54] is a linear-scaling DFT code which uses a similar functional with a numerical atomic orbital basis. This has the advantage over Gaussians that there are no "tails" which need to be neglected. This also allows for very efficient calculations of matrix elements. Another code which uses a general energy functional is QUICKSTEP, which is part of the CP2K package [55]. It uses a hybrid basis of Gaussians and planewaves and is highly suited to large, dense systems such as liquids and solids.

Yang [56] proposed a similar minimization approach with a choice of two energy functionals, where the density matrix is represented in terms of localized orbitals. This method relies on the use of a generalized inverse overlap matrix.

Orbital minimization methods in general work well for larger basis sets but require a good initial guess and take a large number of iterations to converge, as well as having a relatively high crossover point [17].

Other Methods

Other methods include the recursion method of Haydock [57], which uses a recursion scheme to calculate the diagonal elements of the Green's function. This is quite a general method but there can be problems calculating forces. Baroni and Gianozzi [58] also employ a recursion method, where the density is represented on a grid and calculated via a finite difference scheme. This however fails to take into account the smoothness of the density and therefore has a large prefactor. The Green's function

has also been used to calculate bond order potentials which have been reviewed by
Pettifor [59]. Tsuchida and Tsukada [60] used a finite element basis combined with
an unconstrained minimization approach and localization regions.

3.3 Density Matrix Methods

A large number of linear-scaling methods are based around the use of the single
particle density matrix, as this is a fundamental quantity whose localization properties
are well known and therefore such methods allow one to take advantage of the
principle of nearsightedness. This is also the method used in ONETEP and so in the
next section we will focus on DM methods in greater detail than those methods
outlined in the previous section.

3.3.1 The Density Matrix

We begin by defining the density operator $\hat{\rho}$ in terms of the Kohn-Sham single-particle
states:

$$\hat{\rho} = \sum_n f_n |\psi_n\rangle \langle \psi_n|, \tag{3.3}$$

where the f_n are the occupancies. We have assumed the calculation of eigenstates
at the Γ-point only and therefore removed all $\mathbf{k}$-point indices. We also continue to
neglect the effects of spin and so will not be including any factors accounting for
spin degeneracy. The density operator is the projection operator onto the subspace
of occupied states and it has the property of idempotency, such that:

$$\hat{\rho}^2 = \hat{\rho}. \tag{3.4}$$

Furthermore, the DM also must be normalized so that the total number of electrons
N is given by:

$$N = \text{Tr}\,(\rho), \tag{3.5}$$

where ρ is the DM which can be written directly in the position representation as:

$$\rho\left(\mathbf{r}, \mathbf{r}'\right) = \sum_n f_n \psi_n\left(\mathbf{r}\right) \psi_n^*\left(\mathbf{r}'\right). \tag{3.6}$$

Fulfilling these conditions is equivalent to imposing the orthonormality constraint in
standard DFT; it will result in the occupation numbers of the ground state DM being
either one (for those states below the Fermi level) or zero (for those states above the
Fermi level).

For a non-orthogonal basis the density operator can equivalently be written in the following separable form [61]:

$$\hat{\rho} = \sum_{\alpha\beta} |\phi_\alpha\rangle K^{\alpha\beta} \langle\phi_\beta|, \tag{3.7}$$

where $K^{\alpha\beta}$ is known as the density kernel and the $\{|\phi_\alpha\rangle\}$ are a set of orbitals known as "support functions", which are generally non-orthogonal and atom-centred. Due to the non-orthogonality of the orbitals, it is necessary to consider a tensorially correct formulation including the use of a dual basis $\{|\phi^\alpha\rangle\}$ [62, 63], which is defined such that:

$$|\phi^\alpha\rangle = \sum_{\beta} |\phi_\beta\rangle S^{\beta\alpha}, \tag{3.8}$$

where:

$$S^{\alpha\beta} = \langle\phi^\alpha|\phi^\beta\rangle = \left(S_{\alpha\beta}\right)^{-1} \tag{3.9}$$

is the inverse overlap matrix, which is used to convert between covariant and contravariant vectors. The density kernel is therefore the representation of the DM in terms of this dual basis. Additionally, we note that:

$$\langle\phi^\alpha|\phi_\beta\rangle = \delta^\alpha_\beta. \tag{3.10}$$

Within this representation for the DM, the total energy of the non-interacting system, E, is defined as:

$$E = \mathrm{Tr}\,(\mathbf{KH}) \tag{3.11}$$

and the normalization constraint can be re-written as:

$$N = \mathrm{Tr}\,(\mathbf{KS}). \tag{3.12}$$

Finally we note that the density $n\,(\mathbf{r})$ is related to the DM via:

$$n\,(\mathbf{r}) = \rho\,(\mathbf{r}, \mathbf{r}). \tag{3.13}$$

3.3.2 Truncation and Localization

It is possible to solve for the DM directly, however the imposition of the idempotency constraint combined with the fact that the information in the DM already scales quadratically results in no advantage being gained over conventional methods. Therefore, a key concept required to give linear-scaling is the imposition of locality, which is needed to guarantee the sparsity of the DM. This can be done either by thresholding, whereby any matrix elements below a certain value are neglected, or more commonly by the definition of localization regions (LRs), where any matrix elements corresponding to the overlap of orbitals outside these regions are explicitly neglected. These LRs are generally fixed and atom-centred, however they can also be bond-centred or adaptive, following the orbitals during the minimization procedure [64]. The size of the LRs, which are generally spherical and thus defined by their radius, is an important parameter which must be adjusted to compromise between accuracy and efficiency.

3.3.3 Enforcing Idempotency

Another important decision which must be made for such methods is the choice of minimization procedure, which can in general be either constrained or unconstrained. Minimization of the DM without the idempotency constraint would lead to all eigenvalues above the Fermi energy (i.e. unoccupied states) tending to infinity, and those below the Fermi level (i.e. occupied states) tending towards minus infinity, rather than the correct values of zero and one respectively, therefore some constraints are clearly necessary. However, the direct imposition of the idempotency constraint on the DM does not scale linearly, thus an alternative method of enforcing the condition of idempotency must be found. There are a number of methods of doing so which we introduce in this section.

The Purification Transformation

This purification transformation [65] involves an iterative change to the DM such that:

$$\rho_{k+1} = 3\rho_k^2 - 2\rho_k^3, \tag{3.14}$$

where k denotes the iteration number. Note that if ρ is already idempotent, the transformation does not produce any change. For all other ρ, the level of idempotency will consistently increase, providing the initial eigenvalues are between -0.5 and 1.5. However, this range of occupancies only leads to the condition of "weak" idempotency, where the occupation numbers will be between zero and one, not necessarily equal to one. This can lead to instabilities due to flipping between zero and one, therefore to avoid this problem the stricter interval of between $\frac{1-\sqrt{3}}{2}$ and $\frac{1+\sqrt{3}}{2}$ should instead be enforced. The effect of the transformation on the occupation numbers is illustrated by Fig. 3.1.

There are also two variations on the purification method: adaptive purification [66–68], which can be used to return occupancies to within the required range for the usual purification transformation, and canonical purification [69, 70], which can

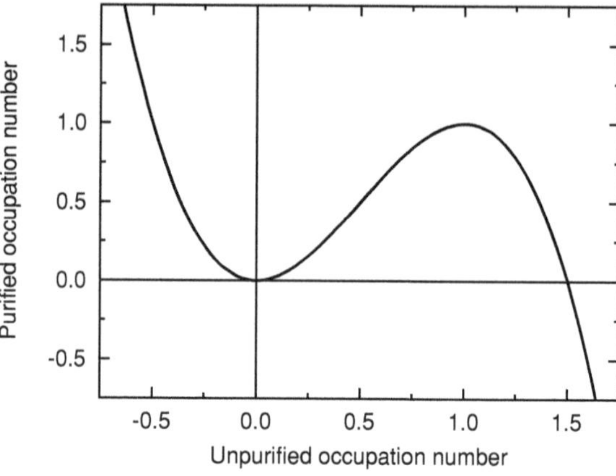

Fig. 3.1 Graph demonstrating the effect of the purification transformation on a given occupation number

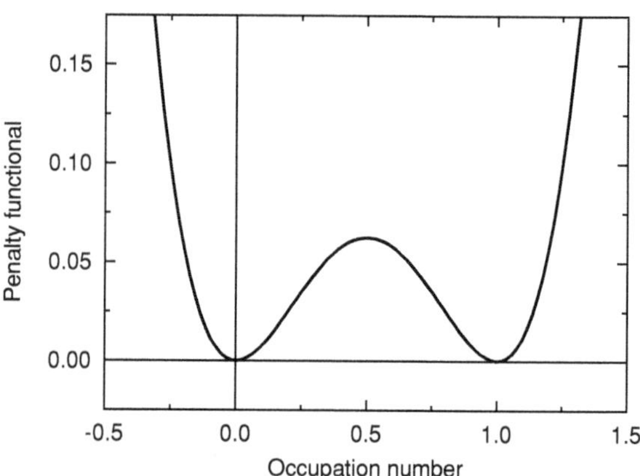

Fig. 3.2 Graph showing the behaviour of the McWeeny's penalty functional for a single occupation number

be used to construct an initial guess for the density kernel in a non self-consistent process.

Penalty Functionals

A related method for imposing idempotency is the use of a penalty-functional [65, 71] such as the one originally proposed by McWeeny:

$$P\left[\rho\right] = \mathrm{Tr}\left[\left(\rho^2 - \rho\right)^2\right] = \sum_n \left(f_n^2 - f_n\right)^2,\tag{3.15}$$

which vanishes only if the DM is idempotent. This is illustrated in Fig. 3.2 for a single occupation number, where it is clear that minima exist at $f_n = 1$ and $f_n = 0$. As it is simply a quartic it is a very easy functional to minimize; one could imagine a scheme where a generalized energy functional $Q[\rho]$ is defined as:

$$Q[\rho] = E[\rho] + \alpha P[\rho] \tag{3.16}$$

and minimized with respect to ρ to impose idempotency. However, only approximate idempotency can ever be achieved in such a manner, due to the variation of the original total energy functional $E[\rho]$ with respect to the orbital occupancies. There can also be problems with multiple local minima, however with a combination of a reasonable initial guess for the DM and the application of normalization constraints, such problems can be avoided.

The LNV Method

One method which uses the purification transformation to drive the DM towards idempotency is that of Li, Nunes and Vanderbilt (LNV) [72, 73], which was independently formulated by Daw [74]. The LNV method involves defining an auxiliary matrix σ, which is related to the DM via the purification transformation, so that:

$$\rho = 3\sigma^2 - 2\sigma^3. \tag{3.17}$$

An auxiliary kernel, $\mathbf{L}$ is also defined, so that:

$$\mathbf{K} = 3\mathbf{LSL} - 2\mathbf{LSLSL}. \tag{3.18}$$

The relationship between the auxiliary DM and auxiliary kernel is the same as that between the actual DM and kernel, i.e.:

$$\sigma\left(\mathbf{r}, \mathbf{r}'\right) = \sum_{\alpha\beta} \phi_\alpha\left(\mathbf{r}\right) L^{\alpha\beta} \phi_\beta^*\left(\mathbf{r}'\right). \tag{3.19}$$

One then proceeds by minimizing the total energy with respect to the auxiliary kernel, so that the DM is naturally driven towards idempotency. As with the purification transformation, the method becomes unstable if the eigenvalues of $\mathbf{L}$ stray below -0.5 or above 1.5, however it is generally a well behaved functional and is guaranteed to provide a variational estimate of the ground state energy. This is a widely used method and has formed the basis of many variations including [75–78].

3.3.4 Example Implementations

CONQUEST [79] is a linear-scaling code which uses a DM implementation based on the LNV method, whereby a spatial cut-off is imposed on the auxiliary DM. The support functions are represented by a choice of B-spline basis functions and pseudo-atomic orbitals. The overall scheme involves a combination of three loops: the inner loop finds the ground state DM with respect to the density kernel for fixed support functions and electron density, the middle loop is used to find self-consistency and the outer loop minimizes the energy with respect to the support functions. This method is known as the optimal basis density minimization method and is related to the method used in ONETEP. Preconditioning is also applied. An early example of an application of CONQUEST was to over 6000 atoms of Si [7]. Another study is of Ge hut clusters on Si [80]; recent calculations have reached system sizes of over one million atoms [81].

3.4 ONETEP

The linear-scaling DFT code which we have used in this work is ONETEP [1–3], which uses a DM formalism of DFT, as outlined in Sect. 3.3 above, therefore avoiding the need to explicitly orthogonalize extended orbitals. As ONETEP is designed for application to large systems, either with large unit cells or using the super-cell approximation, only a single **k**-point need be treated. This is chosen to be the Γ-point, which has the added benefit that the Kohn-Sham eigenstates and therefore the basis set and related quantities can be chosen to be real. The DM is expressed in the separable manner in Eq. 3.7, where the $\{|\phi_\alpha\rangle)\}$ are real, non-orthogonal, localized atom-centred functions. Both the density kernel and support functions are individually optimized, which will be discussed in the following sections, as will the other main features of ONETEP.

3.4.1 NGWF Optimization

Given that one of the key features required for a linear-scaling formulation of DFT is the ability to take advantage of the inherent locality of an electronic system, it is important to use a localized basis set. In ONETEP, this basis set takes the form of non-orthogonal generalized Wannier functions (NGWFs) [82], which are real atom-centred functions that are themselves represented in terms of a basis set of periodic cardinal sine (psinc) functions [82–84], which are discussed further in the following section. This representation means the NGWFs can be optimized during the calculation, enabling a minimal sized basis set which is designed to reflect the chemical environment of the specific system in question and so a high level of accuracy can be maintained. This can be seen from the elimination of basis set superposition errors, which commonly occur in other approaches using localized basis sets [85].

The localization is achieved by the definition of cut-off radii (r_α), outside of which the NGWFs are forced to be zero. This cut-off radius can be made to vary from atom to atom, as can the number of NGWFs situated at each atom, which is

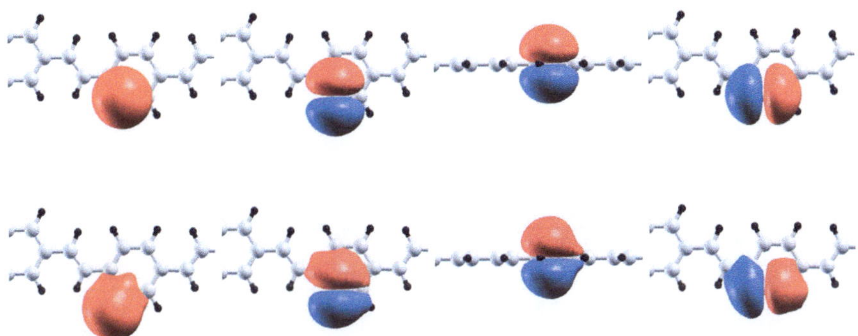

Fig. 3.3 Demonstration of the NGWF optimization process for the polymer poly(*para*-phenylene vinylene). The *top row* shows the initial unoptimized NGWFs on a selected C atom and the *bottom row* shows the final optimized NGWFs for the same atom

usually chosen to reflect the bonding state of the atom in question. To ensure a truly accurate result, the total energy must therefore be converged with respect to both the number and radii of the NGWFs. Alternatively, one could perform a lower accuracy calculation corresponding to a minimal basis set approach by eliminating the NGWF optimization process and instead using the initial guesses generated for the NGWFs, which are generated either by truncating Slater-type contracted Gaussian atomic orbitals or by solving the Kohn-Sham equations for an isolated atom described by a pseudopotential.

The optimization process, which is done in situ, is achieved by minimizing the total energy with respect to the NGWFs, using a preconditioned conjugate gradients scheme. The preconditioning is particularly important as it ensures that the number of iterations required for convergence does not increase with increasing system size, thus enabling true linear-scaling behaviour. An example of the NGWF optimization process is shown in Fig. 3.3.

One additional benefit of the localization of the NGWFs is that empty space which is not covered by them is virtually free from the point of view of computational effort. This means that calculations involving lower dimensional systems are particularly advantageous and so the crossover point at which linear-scaling becomes cheaper than traditional cubic-scaling methods will occur for fewer atoms for such systems.

3.4.2 The Psinc Basis Set

The NGWFs are represented by a localized basis set of psincs, which are ortho-normal and can be related to plane-waves via a Fourier transform, thus allowing a real-space representation and facilitating comparison with the standard plane-wave pseudopotential DFT implementation. Like plane-waves, the basis set quality can be

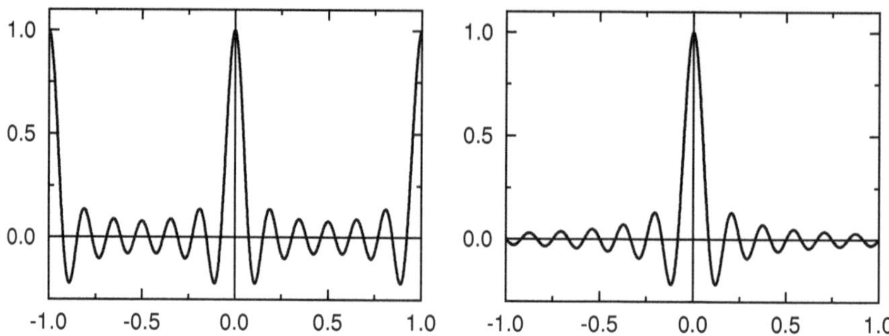

Fig. 3.4 Comparison of a one-dimensional psinc function (*left*) with a sinc function (*right*)

systematically controlled by an energy cut-off parameter, giving a good degree of control over the final level of accuracy.

These psinc functions are periodic bandwidth limited delta functions which can be thought of as a form of periodic sinc function; the similarities between sinc and psinc functions are illustrated by Fig. 3.4. They are distributed evenly over a real space grid such that the function centred at $\mathbf{r}_{klm}$ is written as:

$$D_{klm}\left(\mathbf{r}\right) = D\left(\mathbf{r} - \mathbf{r}_{klm}\right), \tag{3.20}$$

so that $D_{klm}\left(\mathbf{r} - \mathbf{r}_{klm}\right)$ is given by:

$$D_{klm}\left(\mathbf{r} - \mathbf{r}_{klm}\right) = \frac{1}{N_1 N_2 N_3} \sum_{p=-J_1}^{J_1} \sum_{q=-J_2}^{J_2} \sum_{s=-J_3}^{J_3} e^{i(p\mathbf{b}_1+q\mathbf{b}_2+s\mathbf{b}_3)\cdot(\mathbf{r}-\mathbf{r}_{klm})}, \tag{3.21}$$

where the $\{\mathbf{b}_i\}$ are the set of reciprocal lattice vectors and N_i is the number of grid points in the direction of the lattice vector $\mathbf{a}_i$, such that $N_i = 2J_i + 1$. The grid points are then defined by:

$$\mathbf{r}_{klm} = \frac{k}{N_1}\mathbf{a}_1 + \frac{l}{N_2}\mathbf{a}_2 + \frac{m}{N_3}\mathbf{a}_3. \tag{3.22}$$

The NGWFs $\{\phi_\alpha\left(\mathbf{r}\right)\}$ are expressed in this basis by:

$$\phi_\alpha\left(\mathbf{r}\right) = \sum_{k=0}^{N_1-1} \sum_{l=0}^{N_2-1} \sum_{m=0}^{N_3-1} C_{klm,\alpha} D_{klm}\left(\mathbf{r}\right), \tag{3.23}$$

with the $C_{klm,\alpha}$ as the expansion coefficients.

The psinc functions have a number of useful properties, which are demonstrated elsewhere [82, 83] including the fact that they are localized in such a manner that they have a nonzero value only at the grid point at which they are centred. Furthermore, they are orthogonal, which has the useful effect that there are no contributions to the

NGWF gradient outside the localization regions [1], which confers stability on the optimization process. Additionally, one can show that for any cell periodic function $f(\mathbf{r})$, the overlap integral between that function and a psinc basis function is exactly equal to a sum over grid points, i.e.:

$$\int_V f^*(\mathbf{r}) D_{klm}(\mathbf{r}) \, d\mathbf{r} = \int_V f_D^*(\mathbf{r}) D_{klm}(\mathbf{r}) \, d\mathbf{r} \tag{3.24}$$

$$= \Omega \sum_{f=0}^{N_1-1} \sum_{g=0}^{N_2-1} \sum_{h=0}^{N_3-1} f_D^*\left(\mathbf{r}_{fgh}\right) D_{klm}\left(\mathbf{r}_{fgh}\right) = \Omega f_D^*\left(\mathbf{r}_{klm}\right), \tag{3.25}$$

where $f_D(\mathbf{r})$ is the bandwidth limited version of $f(\mathbf{r})$, V is the volume of the unit cell and Ω is the volume per grid point. Consequently, the integral between that function and any function represented in the psinc basis set can also be expressed as a sum over grid points.

The relation between psincs and plane-waves is such that systematic convergence of the number of grid points in the unit cell is equivalent to increasing the number of plane-waves in the PWPP formulation of DFT. The grid spacing can also be directly related to the plane-wave kinetic energy cut-off [86] which means that direct comparisons can be made between ONETEP results and those of the PWPP code CASTEP [87], which can use identical pseudopotentials.

3.4.3 Kernel Optimization

The density kernel in ONETEP is truncated in such a manner that matrix elements between NGWFs situated on atoms which are separated by more than some distance r_k are set to zero. Whilst calculations can be successfully performed without kernel truncation (and indeed no truncation has been applied for the calculations presented in subsequent chapters in this thesis), the imposition of a kernel cut-off is a requirement for true linear-scaling behaviour.

The density kernel optimization process in ONETEP is achieved by a combination of the methods described in Sect. 3.3.3. Firstly, an initial guess for the kernel is constructed using the canonical purification method, after which the approximate penalty method is then used to self-consistently improve the kernel. The optimization process then continues using a variant of the LNV method. During this stage the extremal eigenvalues are tracked to ensure they remain within the required stability range of LNV and if needed adaptive purification is used to return the occupancies to a stable range. The method is variational [88] and due to the combination of different methods used the overall optimization process is very robust.

It is important to apply a normalization constraint during the minimization process, which can be done in a number of ways. The approach used in ONETEP, for reasons of stability when kernel truncation is applied, involves forming a DM which is constructed to be both normalized and purified via the following rescaling expression:

$$\rho = N \frac{3\sigma^2 - 2\sigma^3}{\mathrm{Tr}\left(3\sigma^2 - 2\sigma^3\right)}. \tag{3.26}$$

Further details on density kernel optimization in ONETEP are presented in [89].

As both the NGWFs and the density kernel are truncated in a predefined manner, the Hamiltonian, kernel and overlap matrices will have a predetermined sparsity pattern and so can be multiplied together in order $\mathcal{N}$ operations using efficient sparse matrix schemes [3, 90].

3.4.4 FFT Boxes and True Linear-Scaling

Certain terms in the total energy, such as the kinetic energy term, can be more efficiently calculated in reciprocal space, and so it becomes desirable to use fast Fourier transforms for converting between real and reciprocal space. However, rather than employing FFTs in the usual manner, the FFT box technique [91] is employed; instead of performing an FFT over the entire simulation cell, smaller regions with a fixed radius are associated with each NGWF, within which the FFTs are performed. As the FFT box size relates to the NGWF radii, it becomes independent of system size, thereby avoiding the $\mathcal{O}\left(\mathcal{N}\log\left(\mathcal{N}\right)\right)$ scaling that is normally associated with FFTs.

This approximation is a reasonable one, provided the FFT boxes are sufficiently large compared to the NGWFs, including the NGWF in question and all of its overlapping neighbours. The localization of the NGWFs in real space means that their Fourier transforms are broad in reciprocal space and so the coarser sampling that results from the use of the FFT box compared to whole cell FFTs will have minimal impact on overall accuracy. Furthermore, as the FFT box is defined to be the same shape and size for all NGWFs in the system, the application of operators will be consistent across the system. The size of the FFT box is an important factor in the determination of the crossover point, as a large fraction of computational time is spent on FFTs, as with the PWPP approach. Therefore, the crossover can be expected to occur roughly at the point where the size of the simulation cell has become equal to or greater than the size of the FFT box.

3.4.5 Conduction States in ONETEP

In a standard ONETEP calculation the energy and density are determined from the DM and NGWFs, while the individual eigenstates are not explicitly considered. They can,

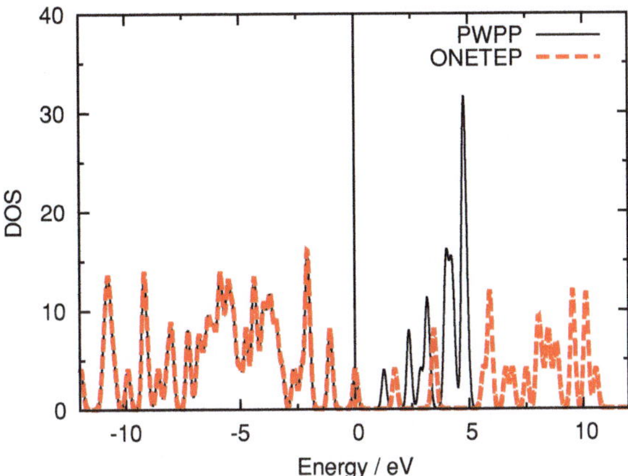

Fig. 3.5 DOS of the conjugated polymer poly(*para*-phenylene vinylene), calculated using both ONETEP and a PWPP code. There is excellent agreement between the two methods for the occupied states, but above the Fermi level the results no longer agree

however, be recovered by a single diagonalization of the Hamiltonian matrix in the basis of NGWFs at the end of a calculation, but only the occupied Kohn-Sham orbitals are accurately represented. This is because the NGWF optimization is solely focussed on minimizing the bandstructure energy of the occupied states, resulting in a basis that does not accurately represent the unoccupied states [86]. In practice some of the lower lying conduction states are close to the correct values, particularly when they are of a similar character to the valence states, however conduction states which are higher in energy are poorly treated and some can be completely absent. An example of this behaviour is shown in Fig. 3.5. Therefore in order to correctly calculate densities of states, band structures and in particular spectra, where matrix elements between valence and conduction states are needed, it becomes necessary to consider the optimization of a second set of NGWFs. Methods for doing so will be discussed in Chap. 6.

It should be noted here that various methods exist for calculating electronic excitation energies using the *GW* method, which avoid the need for explicitly summing over unoccupied states in order to increase computational efficiency [92–95]. Whilst this would appear to invalidate the need for a method of accurately calculating the unoccupied states, it is still necessary to have a complete basis in order to define a projection operator onto the conduction manifold that requires the identity operator. Therefore even with the existence of such approaches it is important to have a method of creating a basis which is able to accurately represent both the occupied and unoccupied states.

3.4.6 Summary

As we have described above, ONETEP is a linear-scaling code which combines the high accuracy of plane-wave calculations via the use of a psinc basis set, with the speed of minimal basis approaches via the use of in-situ optimized, localized NGWFs [8]. This is achieved in a manner which allows for insight into the local chemical environment and is particularly suited for application to lower dimensional systems. Furthermore, an efficient parallelization scheme has been implemented [96] so that computational effort scales approximately linearly with the number of processors, paving the way for calculations on very large systems.

The ONETEP code has been applied to a variety of systems, demonstrating good agreement with the PWPP method. Example applications include crystalline Si, protein-protein interactions and carbon nanotubes in electric fields [97]. Recent work has also been done on improving the efficiency of ONETEP, in particular the sparse matrix algebra, allowing calculations on systems of over 30,000 atoms [3]. A number of developments have also been recently undertaken [98–100] with the aim of expanding the scope of ONETEP. Through this and other work in the field of linear-scaling DFT, it seems likely that the range of problems to which such order $\mathcal{N}$ methods can be applied should continue to increase in the future.

References

1. Skylaris, C.-K., Haynes, P.D., Mostofi, A.A., Payne, M.C.: J. Chem. Phys. **122**(8), 084119 (2005)
2. Haynes, P.D., Skylaris, C.-K., Mostofi, A.A., Payne, M.C.: Phys. Status Solidi B **243**(11), 2489 (2006)
3. Hine, N.D.M., Haynes, P.D., Mostofi, A.A., Skylaris, C.-K., et al.: Comput. Phys. Commun. **180**(7), 1041 (2009)
4. Hung, L., Carter, E.A.: Chem. Phys. Lett. **475**(4–6), 163 (2009)
5. Shimojo, F., Kalia, R.K., Nakano, A., Vashishta, P.: Phys. Rev. B **77**(8), 085103 (2008)
6. Wu, Z., Neaton, J.B., Grossman, J.C.: Phys. Rev. Lett. **100**(24), 246804 (2008)
7. Goringe, C., Hernández, E., Gillan, M., Bush, I.: Comput. Phys. Commun. **102**(1–3), 1 (1997)
8. Skylaris, C.-K., Haynes, P.D.: J. Chem. Phys. **127**(16), 164712 (2007)
9. Sánchez-Portal, D., Ordejón, P., Artacho, E., Soler, J.M.: Int. J. Quantum Chem. **65**(5), 453 (1997)
10. Heady, L., Fernandez-Serra, M., Mancera, R.L., Joyce, S., et al.: J. Med. Chem. **49**(17), 5141 (2006)
11. Ordejón, P., Artacho, E., Soler, J.M.: Phys. Rev. B **53**(16), R10441 (1996)
12. Kohn, W.: Phys. Rev. Lett. **76**(17), 3168 (1996)
13. Prodan, E., Kohn, W.: PNAS **102**(33), 11635 (2005)
14. Ismail-Beigi, S., Arias, T.A.: Phys. Rev. Lett. **82**(10), 2127 (1999)
15. Kohn, W.: Int. J. Quantum Chem. **56**(4), 229 (1995)
16. He, L., Vanderbilt, D.: Phys. Rev. Lett. **86**(23), 5341 (2001)
17. Goedecker, S.: Rev. Mod. Phys. **71**, 1085 (1999)
18. Galli, G.: Curr. Opin. Solid State Mater. Sci. **1**(6), 864 (1996)
19. Bowler, D.R., Miyazaki, T., Gillan, M.J.: J. Phys.: Condens. Matter **14**(11), 2781 (2002)
20. Jordan, D.K., Mazziotti, D.A.: J. Chem. Phys. **122**(8), 084114 (2005)
21. Rudberg, E., Rubensson, E.H.: J. Phys.: Condens. Matter **23**(7), 075502 (2011)
22. Bowler, D.R., Miyazaki, T.: Rep. Prog. Phys. **75**(3), 036503 (2012)
23. Stephan, U., Drabold, D.A.: Phys. Rev. B **57**, 6391 (1998)
24. Artacho, E., Anglada, E., Diéguez, O., Gale, J.D., et al.: J. Phys.: Condens. Matter **20**(6), 064208 (2008)
25. Bowler, D.R., Aoki, M., Goringe, C.M., Horsfield, A.P., et al.: Modell. Simul. Mater. Sci. Eng. **5**, 199 (1997)

26. Horsfield, A.P., Bratkovsky, A.M.: J. Phys.: Condens. Matter **12**, 1 (2000)
27. von Weizsäcker, C.F.: Z. Phys. **96**(7), 431 (1935)
28. Wang, L.-W., Teter, M.P.: Phys. Rev. B **45**(23), 13196 (1992)
29. García-Aldea, D., Alvarellos, J.E.: J. Chem. Phys. **127**(14), 144109 (2007)
30. Pearson, M., Smargiassi, E., Madden, P.A.: J. Phys.: Condens. Matter **5**(19), 3221 (1993)
31. Huang, C., Carter, E.A.: Phys. Rev. B **81**, 045206 (2010)
32. Wang, Y., Stocks, G.M., Shelton, W.A., Nicholson, D.M.C., et al.: Phys. Rev. Lett. **75**(15), 2867 (1995)
33. Abrikosov, I.A., Simak, S.I., Johansson, B., Ruban, A.V., et al.: Phys. Rev. B **56**(15), 9319 (1997)
34. Zeller, R.: Phys. Rev. B **55**(15), 9400 (1997)
35. Zeller, R.: J. Phys.: Condens. Matter **20**(29), 294215 (2008)
36. Yang, W.: Phys. Rev. Lett. **66**(11), 1438 (1991)
37. Kitaura, K., Ikeo, E., Asada, T., Nakano, T., et al.: Chem. Phys. Lett. **313**(3–4), 701 (1999)
38. Fujita, T., Nakano, T., Tanaka, S.: Chem. Phys. Lett. **506**(1–3), 112 (2011)
39. Gadre, S.R., Shirsat, R.N., Limaye, A.C.: J. Phys. Chem. **98**(37), 9165 (1994)
40. Deshmukh, M.M., Bartolotti, L.J., Gadre, S.R.: J. Phys. Chem. A **112**(2), 312 (2008)
41. Stechel, E.B., Williams, A.R., Feibelman, P.J.: Phys. Rev. B **49**(15), 10088 (1994)
42. Goedecker, S.: J. Comp. Phys. **118**(2), 261 (1995)
43. Baer, R., Head-Gordon, M.: Phys. Rev. B **58**(23), 15296 (1998)
44. Krajewski, F.R., Parrinello, M.: Phys. Rev. B **71**(23), 233105 (2005)
45. Krajewski, F.R., Parrinello, M.: Phys. Rev. B **74**, 125107 (2006)
46. Krajewski, F.R., Parrinello, M.: Phys. Rev. B **75**, 235108 (2007)
47. Drabold, D.A., Sankey, O.F.: Phys. Rev. Lett. **70**(23), 3631 (1993)
48. Goedecker, S., Colombo, L.: Phys. Rev. Lett. **73**, 122 (1994)
49. Mauri, F., Galli, G.: Phys. Rev. B **50**(7), 4316 (1994)
50. Shimojo, F., Kalia, R.K., Nakano, A., Vashishta, P.: Comput. Phys. Commun. **140**(3), 303 (2001)
51. Ordejón, P., Drabold, D.A., Martin, R.M., Grumbach, M.P.: Phys. Rev. B **51**(3), 1456 (1995)
52. Mauri, F., Galli, G., Car, R.: Phys. Rev. B **47**(15), 9973 (1993)
53. Kim, J., Mauri, F., Galli, G.: Phys. Rev. B **52**(3), 1640 (1995)
54. Soler, J.M., Artacho, E., Gale, J.D., Garcia, A., et al.: J. Phys.: Condens. Matter **14**(11), 2745 (2002)
55. VandeVondele, J., Krack, M., Mohamed, F., Parrinello, M., et al.: Comput. Phys. Commun. **167**(2), 103 (2005)
56. Yang, W.: Phys. Rev. B **56**, 9294 (1997)
57. Haydock, R.: Solid State Physics, vol. 35. Academic Press, New York (1980)
58. Baroni, S., Giannozzi, P.: Europhys. Lett. **17**(6), 547 (1992)
59. Pettifor, D.G.: Bonding and Structure of Molecules and Solids. Clarendon press, Oxford (1995)
60. Tsuchida, E., Tsukada, M.: J. Phys. Soc. Jpn. **67**(11), 3844 (1998)
61. Hernández, E., Gillan, M.J.: Phys. Rev. B **51**(15), 10157 (1995)
62. White, C.A., Maslen, P., Lee, M.S., Head-Gordon, M.: Chem. Phys. Lett. **276**(1–2), 133 (1997)
63. Artacho, E., Miláns del Bosch, L.: Phys. Rev. A **43**(11), 5770 (1991)
64. Fattebert, J.-L., Gygi, F.: Comput. Phys. Commun. **162**(1), 24 (2004)
65. McWeeny, R.: Rev. Mod. Phys. **32**(2), 335 (1960)
66. Kryachko, E.S.: Chem. Phys. Lett. **318**(1–3), 210 (2000)
67. Holas, A.: Chem. Phys. Lett. **340**(5–6), 552 (2001)
68. Habershon, S., Manby, F.R.: Chem. Phys. Lett. **354**(5–6), 527 (2002)
69. Palser, A.H.R., Manolopoulos, D.E.: Phys. Rev. B **58**, 12704 (1998)
70. Niklasson, A.M.N., Tymczak, C.J., Challacombe, M.: J. Chem. Phys. **118**(19), 8611 (2003)
71. Haynes, P.D., Payne, M.C.: Phys. Rev. B **59**(19), 12173 (1999)
72. Li, X.-P., Nunes, R.W., Vanderbilt, D.: Phys. Rev. B **47**(16), 10891 (1993)

73. Nunes, R.W., Vanderbilt, D.: Phys. Rev. B **50**, 17611 (1994)
74. Daw, M.S.: Phys. Rev. B **47**(16), 10895 (1993)
75. Hernández, E., Gillan, M.J., Goringe, C.M.: Phys. Rev. B **53**, 7147 (1996)
76. Carlsson, A.E.: Phys. Rev. B **51**(20), 13935 (1995)
77. Millam, J.M., Scuseria, G.E.: J. Chem. Phys. **106**(13), 5569 (1997)
78. Challacombe, M.: J. Chem. Phys. **110**(5), 2332 (1999)
79. Bowler, D.R., Choudhury, R., Gillan, M.J., Miyazaki, T.: Phys. Status Solidi B **243**, 989 (2006)
80. Miyazaki, T., Bowler, D.R., Gillan, M.J., Ohno, T.: J. Phys. Soc. Jpn. **77**(12), 123706 (2008)
81. Bowler, D.R., Miyazaki, T.: J. Phys.: Condens. Matter **22**(7), 074207 (2010)
82. Skylaris, C.-K., Mostofi, A.A., Haynes, P.D., Diéguez, O., et al.: Phys. Rev. B **66**(3), 035119 (2002)
83. Mostofi, A.A., Skylaris, C.-K., Haynes, P.D., Payne, M.C.: Comput. Phys. Commun. **147**(3), 788 (2002)
84. Mostofi, A.A., Haynes, P.D., Skylaris, C.-K., Payne, M.C.: J. Chem. Phys. **119**(17), 8842 (2003)
85. Haynes, P.D., Skylaris, C.-K., Mostofi, A.A., Payne, M.C.: Chem. Phys. Lett. **422**(4–6), 345 (2006)
86. Skylaris, C.-K., Haynes, P.D., Mostofi, A.A., Payne, M.C.: J. Phys.: Condens. Matter **17**(37), 5757 (2005)
87. Clark, S.J., Segall, M.D., Pickard, C.J., Hasnip, P.J., et al.: Z. Kristallogr. **220**(5–6), 567 (2005)
88. Skylaris, C.-K., Diéguez, O., Haynes, P.D., Payne, M.C.: Phys. Rev. B **66**, 073103 (2002)
89. Haynes, P.D., Skylaris, C.-K., Mostofi, A.A., Payne, M.C.: J. Phys.: Condens. Matter **20**(29), 294207 (2008)
90. Hine, N.D.M., Haynes, P.D., Mostofi, A.A., Payne, M.C.: J. Chem. Phys. **133**(11), 114111 (2010)
91. Skylaris, C.-K., Mostofi, A.A., Haynes, P.D., Pickard, C.J., et al.: Comput. Phys. Commun. **140**, 315 (2001)
92. Berger, J.A., Reining, L., Sottile, F.: Phys. Rev. B **82**(4), 041103 (2010)
93. Umari, P., Stenuit, G., Baroni, S.: Phys. Rev. B **81**(11), 115104 (2010)
94. Giustino, F., Cohen, M.L., Louie, S.G.: Phys. Rev. B **81**(11), 115105 (2010)
95. Rocca, D., Lu, D., Galli, G.: J. Chem. Phys. **133**(16), 164109 (2010)
96. Skylaris, C.-K., Haynes, P.D., Mostofi, A.A., Payne, M.C.: Phys. Status Solidi B **243**(5), 973 (2006)
97. Skylaris, C.-K., Haynes, P.D., Mostofi, A.A., Payne, M.C.: J. Phys.: Condens. Matter **20**(6), 064209 (2008)
98. Hine, N.D.M., Robinson, M., Haynes, P.D., Skylaris, C.-K., et al.: Phys. Rev. B **83**(19), 195102 (2011)
99. O'Regan, D.D., Hine, N.D.M., Payne, M.C., Mostofi, A.A.: Phys. Rev. B **82**, 081102 (2010)
100. Dziedzic, J., Helal, H.H., Skylaris, C.-K., Mostofi, A.A., et al.: Europhys. Lett. **95**(4), 43001 (2011)

Chapter 4
Theoretical Spectroscopy

Theoretical spectroscopy is a tool of growing importance both in understanding experimental results and making predictions about new materials. The information obtained can help with the interpretation of experimental results, or can be used in tandem with experiment to enable the development of materials with a particular property in mind. Between the different types of spectra which can be measured experimentally, including optical, X-ray, electron energy loss spectra, vibrational and nuclear magnetic resonance (NMR), a wide range of information can be gathered about materials. Indeed, continuing developments in experimental spectroscopy have resulted in a number of improvements, such as the achievement of ever greater spectroscopic resolution. This has allowed for example the study of biological cells using X-ray microscopy on the nano-scale and the ability to control chemical reactions by activating an electronic excitation. See Onida et al. [1] and references within for more examples.

As experimental spectroscopic techniques continue to improve, the field of theoretical spectroscopy becomes increasingly important. Using simulation, it is possible to analyse spectra to a level of detail which is hard to achieve experimentally and thus theoretical spectroscopy calculations are of vital importance in furthering our understanding of experimental results. For example, structures can be modified to determine the effect on spectra and one can also identify which electronic transitions correspond to a particular peak.

This work will focus on the calculation of optical absorption spectra, however this could be extended in future to include other types of spectra involving electronic excitations, for example the calculation of EELS and X-ray absorption spectra (XAS) and so we will also briefly discuss both of these. We will then present the method which we will be using for the calculation of optical spectra, including the derivation of Fermi's golden rule from time-dependent perturbation theory. We will then outline further relevant details for the implementation of this method, including the application within periodic boundary conditions and the use of the scissor operator.

L. Ratcliff, *Optical Absorption Spectra Calculated Using Linear-Scaling Density-Functional Theory*, Springer Theses, DOI: 10.1007/978-3-319-00339-9_4,

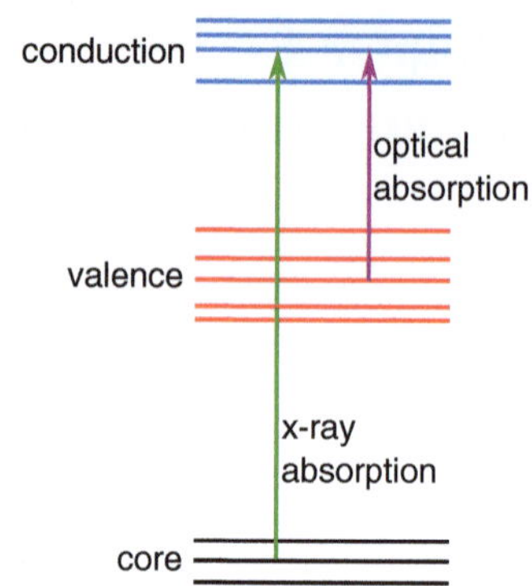

Fig. 4.1 Simple schematic showing example transitions between different electronic energy levels, which are caused by the absorption of a photon either in the X-ray or optical region of the electromagnetic spectrum

4.1 Electronic Excitations

The field of spectroscopy is a vast one and it is beyond the scope of this work to cover all types of spectroscopy. Rather, we are interested here specifically in the types of spectroscopy which result from electronic excitations in a material. Such spectra cover a wide range of the electromagnetic spectrum, depending on the types of electronic states which are involved in the transition, e.g. in XAS transitions occur between core and unoccupied states, whereas in optical absorption spectroscopy transitions are between valence and unoccupied states, as depicted by Fig. 4.1. By varying the energy of the electromagnetic source used to probe the material (or in the case of EELS the energy of the electrons involved), one can therefore gather information concerning both the densities of occupied and unoccupied electronic states. More information on the types of spectroscopy we discuss in the following can be found in [2].

4.1.1 Optical Spectroscopy

Optical spectra can be used to provide a wealth of information about materials, which are gathered using a combination of techniques, including absorption, reflectivity and photoluminescence. We are interested here in optical absorption spectra, where electrons are excited from valence to conduction states, an effect which is caused by the electric field of the incident photon. As well as the ability to study the electronic properties of the system, optical properties are often themselves of direct interest. In particular, a range of technological applications are based on optical excitations, including light emitting diodes (LEDs) and photovoltaics.

Optical spectroscopy also has the particular advantage that time-resolved studies can be done at a higher resolution than with other spectroscopic methods. Another advantage is that optical spectra experiments are non-destructive, which, when combined with the range of accessible time scales makes it a very popular method. For a detailed overview of the uses and techniques involved in optical spectroscopy, see e.g. [3].

4.1.2 Electron Energy Loss Spectroscopy

Electron energy loss spectroscopy is a highly useful spectroscopic tool for studying the structure of electronic systems, in particular, information can be found about unoccupied states. EELS spectra are generated by passing electrons through a sample of material and measuring their resulting loss in energy. Depending on the energies involved, these losses may be due to different processes, including electronic transitions between valence and unoccupied states (low loss) and electronic transitions between core states and unoccupied states (core loss), which includes energy loss near edge structure (ELNES). In addition, phonon and plasmon excitations also play a role. EELS spectra can be measured in conjunction with scanning transmission electron microscopy (STEM), and high energy and spatial resolutions can be achieved, making it a widely applicable spectroscopic technique. The theory, *ab initio* methods for calculating EELS and some example applications are discussed by Keast and Bosman [4] and more information on EELS in general can be found in [5].

4.1.3 X-Ray Spectroscopy

X-ray spectroscopy is a powerful method for exploring the electronic structure of materials; X-ray emission spectra can be used to study occupied states, whilst absorption spectra can be used to study unoccupied states. Significant benefits of the technique include the ability to study all types of solids, the accessibility of a wide energy range, and a relative insensitivity to chemical impurities and other lattice defects, although the interpretation of X-ray spectra can be relatively complex [6]. X-ray absorption near edge spectroscopy (XANES) is closely related to ELNES, although the energy range available using X-rays is necessarily smaller than that using electrons the process can cause less damage. Both methods are particularly useful at giving bonding information. Further information on XAS can be found within [7, 8].

4.2 Calculating Spectra

A number of methods exist for calculating spectra to varying degrees of accuracy and detail, ranging from the use of perturbation theory via Fermi's golden rule, to time-dependent density-functional theory (TDDFT) and many-body perturbation theory (MBPT), as applied within the *GW* approximation, combined with the use of the Bethe-Salpeter equation, to coupled cluster response theory and configuration interaction methods. However we will restrict ourselves to the description of the final approach used in ONETEP, which is based on the use of perturbation theory. We will therefore begin by recalling the central results of both stationary and time-dependent perturbation theory, which will lead on to the derivation of Fermi's golden rule.

4.2.1 Perturbation Theory

One could imagine using a finite difference approach to calculate the derivative of the total energy of a system with respect to some small change in the external potential. This would be very simple to implement, however such an approach can result in the appearance of numerical errors. Furthermore, it would also be computationally expensive. A much better method is therefore the use of perturbation theory, which is based on the idea that a small change in the Hamiltonian of a given system should result in a small change to both the eigenvalues and eigenvectors compared to the original system. This means that instead of having to completely re-calculate the results, a correction term can be found and added to the original solutions.

There are two main types of perturbations that one can consider in quantum mechanics: stationary and time-dependent. We first outline the basic results for stationary perturbations, after which time-dependent perturbations will be presented in more detail, from which we will derive Fermi's golden rule. We continue to work in atomic units.

Stationary Perturbations

Starting from a system for which all the eigenvalues and eigenvectors are known, defined by the equation:

$$\hat{H}_0 | \psi_n^{(0)} \rangle = \mathcal{E}_n^{(0)} | \psi_n^{(0)} \rangle, \tag{4.1}$$

some change is made to the Hamiltonian such that $\hat{H} \rightarrow \hat{H}_0 + \Delta \hat{H}$. Assuming this change is indeed small, the new Hamiltonian can be expressed in terms of a small parameter λ, so that $\hat{H} \rightarrow \hat{H}_0 + \lambda V$. The new eigenvalue equation therefore becomes:

$$\left(\hat{H}_0 + \lambda V \right) | \psi_n \rangle = \mathcal{E}_n | \psi_n \rangle. \tag{4.2}$$

This allows the new energies and eigenstates to be written as the following power series of λ:

$$\mathcal{E}_n = \mathcal{E}_n^{(0)} + \lambda \mathcal{E}_n^{(1)} + \lambda^2 \mathcal{E}_n^{(2)} + \cdots \tag{4.3}$$

$$| \psi_n \rangle = | \psi_n^{(0)} \rangle + \lambda | \psi_n^{(1)} \rangle + \lambda^2 | \psi_n^{(2)} \rangle + \cdots . \tag{4.4}$$

This expansion can then be used to calculate corrections to different orders of λ, by subsituting the above expressions into Eq. 4.2 and equating powers of λ. For the non-degenerate case, the first and second order expressions for the eigenvalues are:

$$\mathcal{E}_n^{(1)} = \lambda \langle \psi_n^{(0)} | V | \psi_n^{(0)} \rangle \tag{4.5}$$

$$\mathcal{E}_n^{(2)} = \lambda^2 \sum_{i \neq n} \frac{\left| \langle \psi_i^{(0)} | V | \psi_n^{(0)} \rangle \right|^2}{\mathcal{E}_n^{(0)} - \mathcal{E}_i^{(0)}}. \tag{4.6}$$

The first order correction to the eigenstate is given by:

$$|\psi_n^{(1)}\rangle = \lambda \sum_{i \neq n} \frac{\langle \psi_i^{(0)} | V | \psi_n^{(0)} \rangle}{\mathcal{E}_n^{(0)} - \mathcal{E}_i^{(0)}} |\psi_i^{(0)}\rangle. \tag{4.7}$$

For degenerate and near-degenerate states, the corrections must be treated differently: this involves the generation of matrix elements between the degenerate eigenstates, i.e. $\langle \psi_n^{(0)} | V | \psi_m^{(0)} \rangle$, then finding the resulting eigenvalues. These eigenvalues are then multiplied by λ to give the first order correction.

Time-Dependent Perturbations

Starting from the same initial system and basic assumptions as for stationary perturbations, correction terms can also be calculated for a time-varying perturbation of the form $\hat{H} \rightarrow \hat{H}_0 + \Delta\hat{H}(t)$. In this case, the TDSE must be used, giving:

$$i\frac{\partial |\psi(t)\rangle}{\partial t} = \left(\hat{H}_0 + \Delta\hat{H}(t) \right) |\psi(t)\rangle. \tag{4.8}$$

The wavefunction can be expressed in terms of the eigenfunctions of the unperturbed system:

$$|\psi(t)\rangle = \sum_n c_n(t) |\psi_n\rangle, \tag{4.9}$$

where the $c_n(t)$ are time-dependent coefficients to be determined.[1] Inserting this expression into TDSE and using the fact that $\langle \psi_n | \psi_m \rangle = \delta_{nm}$, it can be shown that:

$$i\frac{\partial c_m}{\partial t} = c_m(t)\mathcal{E}_m + \sum_n c_n(t) \langle \psi_m | \Delta\hat{H} | \psi_n \rangle. \tag{4.10}$$

---

[1] The superscripts on the eigenfunctions $|\psi_n\rangle$ have been dropped, as from now on they will always be used to represent the eigenfunctions of the original system $|\psi_n^{(0)}\rangle$. Likewise, the original eigenvalues $\mathcal{E}_n^{(0)}$ will be referred to as $\mathcal{E}_n$.

It is useful to express the coefficients $c_n(t)$ in terms of some further set of coefficients $b_n(t)$ multiplied by an exponential factor:

$$c_n(t) = b_n(t) e^{-i\mathcal{E}_n t}. \tag{4.11}$$

By writing the matrix elements $\langle \psi_m | \Delta \hat{H} | \psi_n \rangle$ as ΔH_{mn} and defining:

$$\omega_{mn} = \mathcal{E}_m - \mathcal{E}_n, \tag{4.12}$$

the following expression can be found:

$$i\frac{\partial b_m}{\partial t} = \sum_n b_n(t) \Delta H_{mn} e^{i\omega_{mn} t}. \tag{4.13}$$

The coefficients $b_n(t)$ can be expressed as a series, in an analogous manner to Eqs. 4.5 and 4.6, although we have not explicitly expressed our perturbation in terms of a parameter λ:

$$b_n(t) = b_n^{(0)}(t) + b_n^{(1)}(t) + b_n^{(2)}(t) + \cdots, \tag{4.14}$$

where the zero order term is simply:

$$b_n^{(0)}(t) = \delta_{ni}, \tag{4.15}$$

given an initial state $|\psi(t=0)\rangle = |\psi_i\rangle$. Higher order coefficients are found iteratively, by re-writing Eq. 4.13 as:

$$i\frac{\partial b_m^{(j)}}{\partial t} = \sum_n b_n^{(j-1)}(t) \Delta H_{mn} e^{i\omega_{mn} t}, \tag{4.16}$$

so that $b_n^{(j)}(t)$ can be found by substituting $b_n^{(j-1)}(t)$ into the right hand side of Eq. 4.16 and integrating. For example, the first order coefficients would be:

$$b_f^{(1)}(t) = -i \int_0^t \sum_n \delta_{ni} \Delta H_{fn}(t') e^{i\omega_{fn} t'} dt' \tag{4.17}$$

$$= -i \int_0^t \Delta H_{fi}(t') e^{i\omega_{fi} t'} dt'. \tag{4.18}$$

The above formalism is only valid in the case where the coefficients have not changed greatly from their initial values.

Fermi's Golden Rule

For the specific case of an oscillating perturbation, time-dependent perturbation theory (TDPT) can be used to derive Fermi's golden rule, which can be applied to the calculation of transition probabilities. Defining the perturbation as:

$$\Delta H\left(\mathbf{r}, t\right) = 2\Delta V\left(\mathbf{r}\right) \cos\left(\omega_0 t\right) \tag{4.19}$$

for $t \geq 0$, substituting into Eq. 4.13 and evaluating the integral gives the following expression for the coefficients:

$$b_f\left(t\right) = -\mathrm{i}\left(\frac{\mathrm{e}^{-\mathrm{i}(\omega_{fi} - \omega_0)t} - 1}{-\mathrm{i}\left(\omega_{fi} - \omega_0\right)} + \frac{\mathrm{e}^{-\mathrm{i}(\omega_{fi} + \omega_0)t} - 1}{-\mathrm{i}\left(\omega_{fi} + \omega_0\right)}\right)\Delta V_{fi}. \tag{4.20}$$

This gives very small values for the coefficients, except when $\omega_{fi} \simeq \pm\omega_0$.[2] We shall consider the case where $\omega_{fi} \simeq \omega_0$ and $\mathcal{E}_f > \mathcal{E}_i$; an equivalent approach can be taken for the opposite case. We can now neglect the second term, and the transition probability from some initial state i to a final state f can be shown to be equal to:

$$\left|b_f\left(t\right)\right|^2 = 4\left|\Delta V_{fi}\right|^2 \frac{\sin^2\left[\frac{1}{2}\left(\omega_{fi} - \omega_0\right)t\right]}{\left(\omega_{fi} - \omega_0\right)^2}. \tag{4.21}$$

For increasing t, the function $\dfrac{\sin^2\left[\frac{1}{2}(\omega_{fi} - \omega_0)t\right]}{\left(\omega_{fi} - \omega_0\right)^2}$ tends towards a delta function with the same area of $\frac{\pi t}{2}$, so that:

$$\left|b_f\left(t\right)\right|^2 = \left|\Delta V_{fi}\right|^2 2\pi t\delta\left(\omega_{fi} - \omega_0\right). \tag{4.22}$$

Whilst for long enough time t this will break the assumption that the $b_f\left(t\right)$ are small, it is common practice to overlook this fact, so that Eq. 4.22 can be arranged and expressed as a transition probability per unit time to give:

$$P_{fi} = 2\pi\left|\Delta V_{fi}\right|^2 \delta\left(\mathcal{E}_f - \mathcal{E}_i - \omega_0\right), \tag{4.23}$$

which is known as Fermi's golden rule.

We therefore have an expression giving the probability of a particular electronic transition, involving a joint density of states between valence and conduction states (the δ term), which is weighted by the appropriate matrix elements. The matrix elements act as selection rules that identify which transitions are allowed or forbidden,

---

[2] This can be considered as being equivalent to the cases of absorption ($\omega_{fi} \simeq \omega_0$) and stimulated emission ($\omega_{fi} \simeq -\omega_0$).

with appropriate weightings for those which are permitted, and the DOS term ensures that energy is conserved during the transition.

Applications of Perturbation Theory

As outlined in the previous Sections, perturbation theory is a very general method which is not limited to just the calculation of experimental spectra; rather it can be applied to a wide range of problems, including lattice dynamics, elastic constants, dielectric constants and the calculation of polarizations and effective charges. Furthermore, it is not a DFT-specific method, and therefore can also be implemented within other electronic structure methods, such as Hartree Fock. The general perturbation can be of any form, however two of the more common are displacements of atoms and the application of electric fields, which can be either static or time dependent and therefore make use of the appropriate version of perturbation theory.

For many applications, the first order derivatives only are needed; however for some applications such as phase transitions, higher orders are needed. These can be found efficiently using the $2n + 1$ theorem, and other methods based on it. See for example Gonze and Vigneron for further discussion [9].

In this work, however, we are interested in how perturbation theory can be applied within DFT to the calculation of experimental spectra, in particular optical absorption spectra. Therefore, we shall focus on the use of perturbation theory within this context. The interested reader should see e.g. [10–14] for methods relating to, amongst other topics, the calculation of vibrational modes, Raman spectra and nuclear magnetic resonance.

4.2.2 Application to Optical Spectroscopy

Following on from Sect. 4.2.1 we shall discuss the calculation of optical absorption spectra and other optical properties following the method [15] applied in the PWPP DFT code CASTEP [16]. The starting point for this is Fermi's golden rule (Eq. 4.23) for calculating the probability of a given electronic energy transition, which naturally divides into two parts: the calculation of matrix elements, and the calculation of a density of states term. The latter is relatively straightforward, and indeed the capability of calculating DOS in ONETEP already exists. The only real complication comes in finding an efficient method of accurately integrating over the entire Brillouin zone, a topic which has been thoroughly discussed by e.g. Pickard [15]. Whilst the conduction states are not automatically well represented in ONETEP, for the purposes of this discussion, we will assume the existence of an accurate DOS; the subject of calculating the DOS accurately in the conduction regime will be returned to in Chap. 6. Although the details of the calculation of a given spectrum naturally depend on the type of spectroscopy involved, the overall procedure for the calculation of XAS and EELS will be similar to that described below.

The Dipole Approximation

The general form of the matrix elements required for Fermi's golden rule is:

$$\Delta V_{fi} = \langle \psi_f | e^{i\mathbf{q}\cdot\mathbf{r}} | \psi_i \rangle, \tag{4.24}$$

where $\mathbf{q}$ takes a slightly different meaning depending on whether a photon or an electron is providing the energy required for the transition, and additional constants are present in the calculation of XAS. Provided $\mathbf{q}$ is small, which will be the case for long wavelengths, this can be expanded as a Taylor series, giving:

$$\langle \psi_f | e^{i\mathbf{q}\cdot\mathbf{r}} | \psi_i \rangle = \langle \psi_f | \psi_i \rangle + i\langle \psi_f | \mathbf{q} \cdot \mathbf{r} | \psi_i \rangle + \frac{1}{2} \langle \psi_f | (\mathbf{q} \cdot \mathbf{r})^2 | \psi_i \rangle + \cdots . \tag{4.25}$$

The $\{|\psi_n\rangle\}$ are mutually orthogonal so the first term becomes zero and in the long-wavelength limit the higher order terms can be neglected, leaving just the dipole term. This is a widely used simplification which is referred to as the dipole approximation, and is a good approximation for optical wavelengths.

The Dielectric Function

We employ Fermi's golden rule and the dipole approximation, using the transition matrix elements to calculate the imaginary part of the dielectric function, $\varepsilon_2 (\omega)$. The dielectric function describes the response of a given material to an external electric field, in particular the imaginary component describes the energy losses which occur in a medium due to electronic transitions and is therefore related to absorption. The connection between the dielectric function and other optical properties is described later in this Section. An approximate form for the imaginary component, based on Fermi's golden rule, is written in atomic units as:

$$\varepsilon_2 (\omega) = \frac{8\pi^2}{\Omega} \sum_{\mathbf{k},v,c} \left| \langle \psi_{\mathbf{k}}^c | \hat{\mathbf{q}} \cdot \mathbf{r} | \psi_{\mathbf{k}}^v \rangle \right|^2 \delta \left(\mathcal{E}_{\mathbf{k}}^c - \mathcal{E}_{\mathbf{k}}^v - \omega \right), \tag{4.26}$$

where v and c denote valence and conduction bands respectively, $|\psi_{\mathbf{k}}^n\rangle$ is the nth eigenstate at a given $\mathbf{k}$-point with a corresponding energy $\mathcal{E}_{\mathbf{k}}^n$, Ω is the cell volume, $\hat{\mathbf{q}}$ is the direction of polarization of the photon and ω its angular frequency.

From the imaginary part of the dielectric function one can then also calculate the real part, $\varepsilon_1 (\omega)$, using the appropriate Kramers-Kronig relation:

$$\varepsilon_1 (\omega_0) = \varepsilon_\infty + \frac{2}{\pi} P \int_0^\infty \frac{\omega \varepsilon_2 (\omega)}{\omega^2 - \omega_0^2} d\omega, \tag{4.27}$$

where P indicates the principal value of the integral. The full complex dielectric function is then given by:

$$\varepsilon(\omega) = \varepsilon_1(\omega) + i\varepsilon_2(\omega). \tag{4.28}$$

Position Versus Momentum

When one applies the dipole approximation, the calculation of transition matrix elements becomes equivalent to the calculation of position matrix elements. However, when periodic boundary conditions are applied, as in this work, the position operator is known to be undefined and it becomes necessary to use the momentum operator. Momentum matrix elements can be easily related to position matrix elements by considering the commutator with the Hamiltonian, but when non-local pseudopotentials are being used, one must be careful to include the commutator between the position operator and the non-local potential. The relation is thus written [17]:

$$\langle \psi_f | \mathbf{r} | \psi_i \rangle = \frac{1}{i\omega_{fi}} \langle \psi_f | \mathbf{p} | \psi_i \rangle + \frac{1}{\omega_{fi}} \langle \psi_f | \left[\hat{V}_{nl}, \mathbf{r} \right] | \psi_i \rangle. \tag{4.29}$$

In the case of core level spectra where the initial states are highly localized, it is possible to make an even more drastic approximation and calculate an angular momentum projected DOS, rather than building a spectrum by explicitly calculating matrix elements with the core wavefunctions. This approach has been used to give qualitatively accurate results, and is discussed by Gao et al. and references within [18].

k-Point Summation

Once the appropriate matrix elements have been found and the probability of a given transition calculated, generating a spectrum is then a straightforward process, which simply requires a sum over all possible initial and final states, to give a weighted DOS, as in Eq. 4.26. In principle this includes a **k**-point sum over the entire Brillouin zone, however as with ground state ONETEP calculations, it is assumed that a large enough supercell will be used such that only the Γ point need be considered. This could be extended in future using methods for interpolating band structures in ONETEP that are discussed in Sect. 5.2. For the purposes of this work, however, all calculations have been restricted to the Γ point only and thus we have dropped the index for **k**-points. For the case of optical spectra, additional optical properties may also be calculated, and for core level spectra, further complications arise with the calculation of core states, in particular core hole screening effects.

The Scissor Operator

For the purposes of comparison with experiment, it is sometimes desirable to make use of the scissor operator, whereby the conduction band energies are rigidly shifted upwards such that the DFT Kohn-Sham band gap is equal to experimental values. Whilst this is not an *ab initio* correction, in practice relatively good agreement can be found with experiment in this manner for many systems without the need for more computationally intensive methods, such as the *GW* approximation, as discussed in Sect. 2.2.4. There will, however, be a number of occasions when it becomes necessary to use less approximate methods.

Improvements to the Method

The approach described above is only an approximate method for the calculation of optical absorption spectra; in addition to the inherent approximations in DFT and the difference between the Kohn-Sham eigenvalues and the true quasiparticle energies, we have also neglected local-field effects and ignored the electron-hole interaction. There are two main methods which improve upon the accuracy: TDDFT and the *GW* approximation. In addition, local-field effects can be included by calculating the full dielectric matrix, rather than just the diagonal elements [19–22].

TDDFT [23] is a generalized version of DFT which involves the re-derivation of the Kohn-Sham equations to allow for a time-dependent potential. This relies on the underlying principle that the time-dependent potential acting on a many-electron system is uniquely determined by the time evolution of the one-electron density for any initial state. It can be applied to the calculation of optical absorption spectra, e.g. [24–26] but suffers with complications when applied to solids. For further details see [27–30].

The *GW* approximation is based on the use of the single-particle Green's function and allows for more accurate calculation of the quasiparticle energies [31–33]. This can be combined with the use of the Bethe-Salpeter equation [34–36], which takes a two-body approach to the calculation of neutral excitations in order to correctly take into account the electron-hole interaction. A number of similar approaches have been taken by various groups who have demonstrated good agreement with experiment [37–40].

The *GW* approximation is computationally more expensive than TDDFT but also gives more accurate results; for a thorough comparison see Onida et al. [1]. However, more accurate schemes such as these are rather more computationally expensive than our method, and as we are aiming at application to large systems, we have necessarily restricted ourselves to a more basic approach. We note that a number of methods have recently been developed for decreasing the cost of the *GW* approximation e.g. that of Umari et al. [41] who used Wannier-like orbitals to improve the applicability of *GW* to larger systems, and so it is anticipated that the implementation in ONETEP could be extended in future.

Other Optical Properties

Once the full dielectric function is known, a number of other optical properties can also be calculated. This includes the complex refractive index, $N(\omega)$, which is related directly to the complex dielectric function via:

$$N(\omega) = \sqrt{\varepsilon(\omega)}. \tag{4.30}$$

Given that N is written as:

$$N(\omega) = n(\omega) + i\kappa(\omega), \tag{4.31}$$

the real and imaginary components of both the refractive index and the dielectric function are therefore related by:

$$\varepsilon_1 = n^2 - \kappa^2 \tag{4.32}$$

$$\varepsilon_2 = 2n\kappa. \tag{4.33}$$

The components of the refractive index can then be used to define both the absorption coefficient, $\alpha(\omega)$:

$$\alpha(\omega) = \frac{2\kappa\omega}{c} \tag{4.34}$$

and the reflection coefficient $R(\omega)$:

$$R(\omega) = \left|\frac{1-N}{1+N}\right|^2 = \frac{(n-1)^2 + \kappa^2}{(n+1)^2 + \kappa^2}. \tag{4.35}$$

Finally, one can define the energy loss function as:

$$\mathrm{Im}\left\{\frac{-1}{\varepsilon(\omega)}\right\}. \tag{4.36}$$

For high energies this becomes approximately equal to the imaginary component of the dielectric function.

From these relations it is clear that the dielectric function plays a key role, in that it can be related to a range of other optical properties. This means that the dielectric function can be extracted from experimental results even when other quantities, such as the absorption and reflection coefficients, which might be easier to determine experimentally, have been directly measured. See e.g. [2, 15] for more information on optical properties of materials.

4.3 Summary

In this chapter we have introduced the topic of spectroscopy, in particular those types of spectroscopy which are caused by electronic excitations, and have highlighted the importance of theoretical spectroscopy in the interpretation of experimental results. We have also described the method we will be using to calculate optical absorption spectra, and given a brief discussion of alternative methods. In Chap. 6 we will present further details relating specifically to the implementation in ONETEP.

References

1. Onida, G., Reining, L., Rubio, A.: Rev. Mod. Phys. **74**(2), 601 (2002)
2. Kuzmany, H.: Solid-State Spectroscopy: an Introduction. Berlin, Springer (2009)
3. Tkachenko, N.: Optical Spectroscopy: Methods and Instrumentations. Elsevier, Boston (2006)
4. Keast, V., Bosman, M.: Mater. Sci. Technol. **24**, 651 (2008)
5. Egerton, R.F.: Electron Energy-Loss Spectroscopy in the Electron Microscope. Language of science. Plenum Press, New York (1996)
6. Parratt, L.G.: Rev. Mod. Phys. **31**(3), 616 (1959)
7. Que, L.: Physical Methods in Bioinorganic Chemistry: Spectroscopy and Magnetism. University Science Books, Sausalito (2000)
8. Rehr, J.J., Kas, J.J., Prange, M.P., Sorini, A.P., et al.: C.R. Phys. **10**(6), 548 (2009)
9. Gonze, X., Vigneron, J.-P.: Phys. Rev. B **39**(18), 13120 (1989)
10. Gonze, X.: Phys. Rev. A **52**(2), 1096 (1995)
11. Baroni, S., Giannozzi, P., Testa, A.: Phys. Rev. Lett. **58**(18), 1861 (1987)
12. Baroni, S., de Gironcoli, S., Dal Corso, A., Giannozzi, P.: Rev. Mod. Phys. **73**(2), 515 (2001)
13. Putrino, A., Sebastiani, D., Parrinello, M.: J. Chem. Phys. **113**(17), 7102 (2000)
14. Refson, K., Tulip, P.R., Clark, S.J.: Phys. Rev. B **73**(15), 155114 (2006)
15. Pickard, C.J.: Ab initio Electron Energy Loss Spectroscopy. Ph.D. thesis, University of Cambridge (1997)
16. Clark, S.J., Segall, M.D., Pickard, C.J., Hasnip, P.J., et al.: Z. Kristallogr. **220**(5–6), 567 (2005)
17. Read, A.J., Needs, R.J.: Phys. Rev. B **44**(23), 13071 (1991)
18. Gao, S.-P., Pickard, C.J., Perlov, A.: J. Phys.: Condens. Matter **21**(10), 104203 (2009)
19. Adler, S.L.: Phys. Rev. **126**, 413 (1962)
20. Wiser, N.: Phys. Rev. **129**, 62 (1963)
21. Baroni, S., Resta, R.: Phys. Rev. B **33**, 7017 (1986)
22. Hybertsen, M.S., Louie, S.G.: Phys. Rev. B **35**, 5585 (1987)
23. Runge, E., Gross, E.K.U.: Phys. Rev. Lett. **52**(12), 997 (1984)
24. Varsano, D., Di Felice, R., Marques, M.A.L., Rubio, A.: J. Phys. Chem. B **110**(14), 7129 (2006)
25. Walker, B., Saitta, A.M., Gebauer, R., Baroni, S.: Phys. Rev. Lett. **96**(11), 113001 (2006)
26. Vasiliev, I., Öğüt, S., Chelikowsky, J.R.: Phys. Rev. B **65**(11), 115416 (2002)
27. Bauernschmitt, R., Ahlrichs, R.: Chem. Phys. Lett. **256**(4–5), 454 (1996)
28. Baer, R.: J. Mol. Struct. (Theochem) **914**(1–3), 19 (2009)
29. Sottile, F., Bruneval, F., Marinopoulos, A.G., Dash, L.K., et al.: Int. J. Quantum Chem. **102**(5), 684 (2005)
30. Reining, L., Olevano, V., Rubio, A., Onida, G.: Phys. Rev. Lett. **88**(6), 066404 (2002)
31. Hedin, L.: Phys. Rev. **139**(3A), A796 (1965)
32. Godby, R.W., Needs, R.J.: Phys. Scr. **T31**, 227 (1990)
33. Del Sole, R., Reining, L., Godby, R.W.: Phys. Rev. B **49**(12), 8024 (1994)

34. Salpeter, E.E., Bethe, H.A.: Phys. Rev. **84**(6), 1232 (1951)
35. Strinati, G.: Phys. Rev. Lett. **49**, 1519 (1982)
36. Nakanishi, N.: Prog. Theor. Phys. Suppl. **43**, 1 (1969)
37. Rohlfing, M., Louie, S.G.: Phys. Rev. B **62**(8), 4927 (2000)
38. Onida, G., Reining, L., Godby, R.W., Del Sole, R., et al.: Phys. Rev. Lett. **75**, 818 (1995)
39. Benedict, L.X., Shirley, E.L., Bohn, R.B.: Phys. Rev. Lett. **80**, 4514 (1998)
40. van der Horst, J.-W., Bobbert, P.A., Michels, M.A.J., Brocks, G., et al.: Phys. Rev. Lett. **83**, 4413 (1999)
41. Umari, P., Stenuit, G., Baroni, S.: Phys. Rev. B **79**, 201104 (2009)

Chapter 5
Basis Sets and Band Structures

In order to evaluate the strengths and weaknesses of the different methods for calculating conduction states in ONETEP, a toy model was created within which three methods were implemented. This toy model was aimed at imitating the main components of ONETEP whilst remaining as simple as possible. The existence of a simple model also allowed the comparison of two different methods of calculating band structures, in particular their strengths and weaknesses when applied within an imperfect basis set. Using this simple model it was possible to identify some of the problems caused by "bad" basis sets, which in ONETEP will correspond to NGWFs which have not been sufficiently well optimized, an inadequate number of NGWFs, or NGWFs with radii which are not big enough.

To this end, we have created a number of localized basis sets, restricting ourselves to the simplest case of nearest neighbour overlap. Whilst this proved to be insufficient for the accurate re-creation of band structures for simple one-dimensional potentials, interesting insights were gained into the requirements of a good basis set. The less strictly localized B-spline functions were used as an example of a "good" basis set.

Within this chapter we will therefore introduce the different localized basis sets, and compare the results for one-dimensional band structures based on two different potentials. We will also present the two methods for band structure calculation and identify cases for which one method might be preferable over another.

5.1 Localized Basis Sets

As previously discussed, one of the central concepts in linear-scaling DFT is the use of a localized basis set, as this enables the exploitation of the underlying nearsightedness of electronic systems. We have therefore considered a number of simple model basis sets which are as localized as possible. As discussed in Sect. 3.3, it is important to maintain a tensorially correct form when nonorthogonal basis sets are used. The wavefunction $|\psi_n\rangle$ for a given state n will therefore be written as:

L. Ratcliff, *Optical Absorption Spectra Calculated Using Linear-Scaling Density-Functional Theory*, Springer Theses, DOI: 10.1007/978-3-319-00339-9_5,
© Springer International Publishing Switzerland 2013

$$|\psi_n\rangle = c_n^\alpha |\phi_\alpha\rangle \tag{5.1}$$

in terms of the basis functions $\{|\phi_\alpha\rangle\}$ and a set of coefficients $\{c_n^\alpha\}$, and we will be solving the generalized Schrödinger equation:

$$H_{\beta\alpha} c_n^\alpha = \mathcal{E}_n S_{\beta\gamma} c_n^\gamma, \tag{5.2}$$

where $\mathbf{H}$ and $\mathbf{S}$ are the Hamiltonian and overlap matrices respectively and we have used the Einstein summation convention for Greek indices. As with the rest of this work, we apply periodic boundary conditions and use atomic units throughout this chapter.

5.1.1 Nearest Neighbour Basis Sets

Basis Set Definitions

We have considered five one-dimensional basis sets which are localized to nearest neighbour overlap only. They are each constructed from a single function $\phi_0(x)$, which is translated by an integer multiple of some length x_0 to find the basis functions $\phi_\alpha(x)$, such that:

$$\phi_\alpha(x) = \phi_0(x - \alpha x_0), \tag{5.3}$$

where $x_0 = L/N$, N being the number of basis functions in the system and L being the length of the unit cell. The matrix elements are calculated using the following definitions for the overlap $S_{\alpha\beta}$, momentum $P_{\alpha\beta}$ and kinetic energy $T_{\alpha\beta}$ matrix elements:

$$S_{\alpha\beta} = \int_{-\infty}^{\infty} \phi_\alpha(x)\, \phi_\beta(x)\, \mathrm{d}x \tag{5.4}$$

$$P_{\alpha\beta} = \mathrm{i} \int_{-\infty}^{\infty} \phi_\alpha(x)\, \frac{\mathrm{d}\phi_\beta(x)}{\mathrm{d}x}\, \mathrm{d}x \tag{5.5}$$

$$T_{\alpha\beta} = -\frac{1}{2} \int_{-\infty}^{\infty} \phi_\alpha(x)\, \frac{\mathrm{d}^2\phi_\beta(x)}{\mathrm{d}x^2}\, \mathrm{d}x. \tag{5.6}$$

As they only overlap with nearest neighbours and the function at each site is simply the translation of that at an adjacent site, we need only consider S_0, S_1, T_0, T_1, P_0 and P_1, where the suffix "0" denotes the integral over a given function with itself, and the suffix "1" denotes the integral over a function with its nearest neighbour. For uniformly distributed basis functions these matrix elements will be independent of the site under consideration.

As well as the five nearest neighbour basis sets, B-splines [1] were also studied for comparison; they have an overlap up to and including third nearest neighbours,

Table 5.1 Basis set definitions and matrix elements for the nearest neighbour basis sets, where the basis functions are written in the form $\phi_\alpha(x) = \phi_0(x - \alpha x_0)$ with x_0 being some lattice width

Name	Definition of $\phi_0(x)$	S_1	$T_0 x_0^2$	$T_1 x_0^2$	$\frac{P_1 x_0}{i}$
Orthog	$(x^2 - x_0^2)(x^2 - \lambda^2 x_0^2) H(x_0^2 - x^2)$	0	2.83	-1.23	0.40
Cusp free	$(x^2 - x_0^2)^2 H(x_0^2 - x^2)$	$\frac{103}{512}$	$\frac{3}{2}$	$-\frac{93}{128}$	$\frac{369}{512}$
Extra cusps	$\left[\left(-\lvert x_0\rvert + \frac{3x_0}{4}\right)^2 - \frac{x_0^2}{16}\right]$ if $0 \le \lvert x\rvert \le x_0$	$-\frac{1}{8}$	$\frac{35}{4}$	$\frac{155}{32}$	$-\frac{5}{8}$
Orthog splines	$\left[1 - \frac{210x^2}{13x_0^2} + \frac{280\lvert x\rvert^3}{13x_0^3}\right]$ if $0 \le \lvert x\rvert \le \frac{x_0}{2}$ $\left[\frac{18}{13} - \frac{108\lvert x\rvert}{13a} + \frac{162x^2}{13a^2} - \frac{72\lvert x\rvert^3}{13x_0^3}\right]$ if $\frac{x_0}{2} \le \lvert x\rvert \le x_0$	0	24.97	6.02	0.93
Mirrored	$\left[1 - \frac{2x^2}{x_0^2}\right]$ if $0 \le \lvert x\rvert \le \frac{x_0}{2}$ $\left[2\left(1 - \frac{2\lvert x\rvert}{x_0} + \frac{x^2}{x_0^2}\right)\right]$ if $\frac{x_0}{2} \le \lvert x\rvert \le x_0$	$\frac{7}{46}$	$\frac{40}{23}$	$-\frac{20}{23}$	$\frac{15}{23}$

$H(x)$ is the Heaviside function. The normalization constant c has been omitted. For the orthogonal basis $\lambda = 0.716$; the cusp free basis has the same form but with $\lambda = 1$. Note P_0 is always zero as the functions are all symmetric, and the matrix elements are normalized such that $S_0 = 1$

and hence are less localized but are known to give accurate results. They are defined as follows, where c is a normalization constant:

$$\phi_0(x) = \begin{cases} c\left[1 - \frac{3}{2}\left(\frac{x}{x_0}\right)^2 + \frac{3}{4}\left(\frac{\lvert x\rvert}{x_0}\right)^3\right] & \text{if } 0 \le \lvert x\rvert \le x_0 \\ c\left[\frac{1}{4}\left(2 - \frac{\lvert x\rvert}{x_0}\right)^3\right] & \text{if } x_0 \le \lvert x\rvert \le 2x_0 \end{cases}.$$

The function definitions and matrix elements for each of the five basis sets studied are given in Table 5.1. They are shown graphically in Fig. 5.1, including B-splines, and the sum over basis functions is also depicted. This can give an indication of how accurately the functions are likely to be able to represent the eigenfunctions of a given system, where those basis sets which sum to a constant, i.e. the B-splines and the mirrored basis set (which was specifically constructed to do so), are more likely to better represent a plane-wave for example, and therefore might be expected to give more accurate results. On the other hand, the basis sets which have very defined cusps would be less likely to accurately represent a plane-wave, but might be better suited to other systems, for example those which are very tightly bound.

Results

Once the appropriate matrix elements have been calculated, the Hamiltonian and overlap matrices can be constructed. Initially a zero potential system was considered, so that the Hamiltonian was just the kinetic energy matrix. The resulting generalized eigenvalue equation was then solved by direct diagonalization.

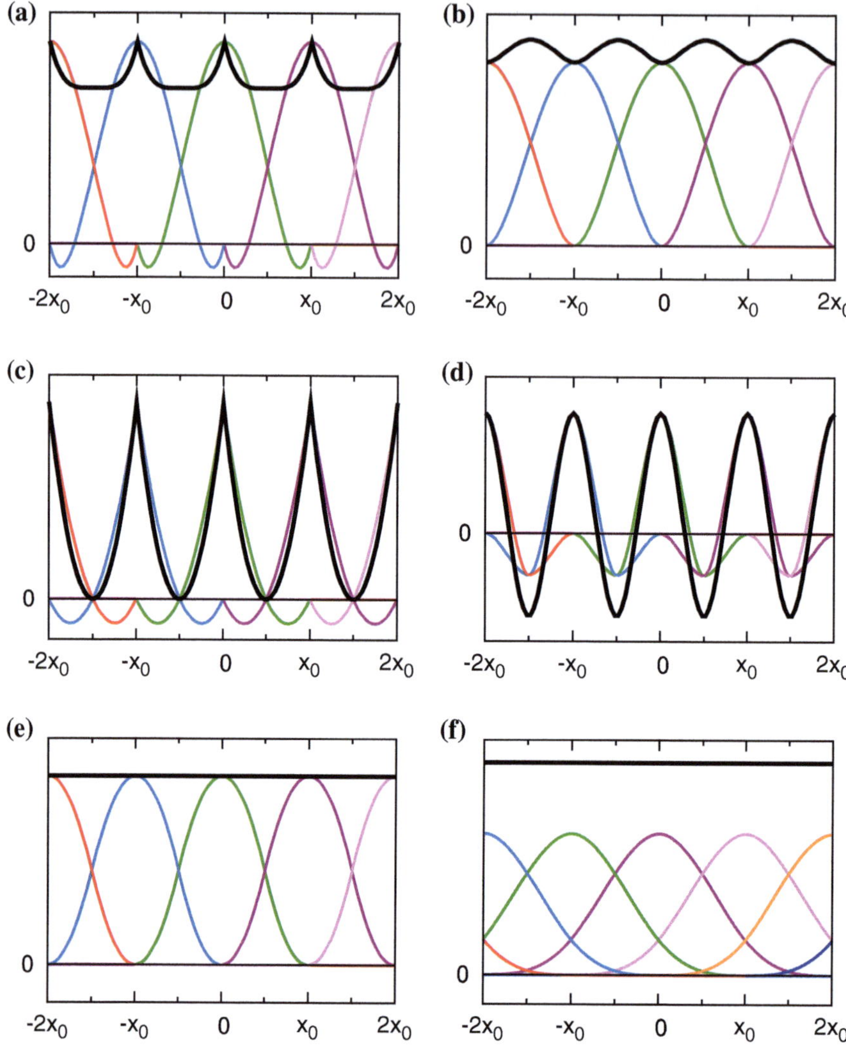

Fig. 5.1 Basis functions and the sum over basis functions for the five nearest neighbour basis sets and B-splines. The basis functions are shown in colour and the sum over functions is shown in *black*. **a** Orthogonal. **b** Cusp free. **c** Extrx cusps. **d** Orthogonal cusps. **e** Mirrored. **f** B-splines

A system size of length $L = 1$ Bohr was used, and in each case $N = 400$ basis functions was deemed sufficient to reach the required level of convergence. The resulting eigenvalues were then compared to the correct values, for which a summary is given in Table 5.2 of selected eigenvalues for each basis set. To facilitate easier comparison with the true results, the eigenvalues are shifted by some fixed amount so that the lowest eigenvalue equals zero. The B-spline results are omitted,

Table 5.2 Eigenvalues for selected low energy bands compared to the correct values calculated for $N = 400$.

Band	Correct values	Orthog	Cusp free	Extra cusps	Orthog splines	Mirrored
2	19.74	48.46	20.64	150.02	237.73	26.32
4	78.96	193.81	82.58	600.08	950.85	105.28
6	177.65	436.02	185.82	1350.21	2139.18	236.89
Correction	N/A	60576	5348	-120000	2067712	0

These have been adjusted to make the first eigenvalue zero, and the size of the applied correction is also shown. For all the bands shown, the B-spline values agreed with the correct values to within 2 dp

as they agree with the correct results to within the level of precision shown. The same trends were observed when testing with a non-periodic system, but with much slower convergence.

Of all the basis sets, only the B-splines give the correct free-electron eigenvalues, as they were expected to. Amongst the nearest neighbour basis sets, two show the greatest agreement with the expected results; the mirrored basis is the only one to have the lowest band pass through zero and so requires no additional shifting of eigenvalues, and the cusp-free basis has the closest values to the true eigenvalues once a shift is applied. All the other basis sets give band structures that are shifted from zero by, in some cases, very large energies, and the spacing of the bands vastly disagrees with the expected results, with increasing errors as the band index increases.

5.1.2 The Requirements for a Good Basis Set

The Effect of Cusps

The results for the different basis sets have been compared with the aim of determining the features that are required for a "good" basis set, i.e. one which gives accurate results without requiring an excessively large number of basis functions. To this end, it is interesting to compare the orthogonal basis set with the cusp-free variation, as they both have the same definition and differ only in the value of the parameter λ which is chosen. The cusp-free variation gives much better results, which implies that the existence of cusps is undesirable. This is due to the discontinuity in the first derivative, which will become a Dirac-δ function in the second derivative and thus result in very large terms in the kinetic energy. This observation is further supported by the fact that the basis set with extra cusps is even further from the expected results. Therefore, as predicted by their ability to represent a constant, cusps are undesirable, at least for the zero potential system considered here. This effect is particularly relevant to ONETEP, as the NGWF optimization process can result in the appearance of kinks at the edge of the localization regions. It is therefore likely that the presence of these kinks will result in decreased accuracy in the energy eigenvalues calculated in such an NGWF basis.

This effect of cusps can be better understood by considering the Fourier transform of a function:

$$\tilde{f}(k) = \frac{1}{\sqrt{2\pi}} \int_{-\infty}^{\infty} f(x)\, e^{ikx} dx. \tag{5.7}$$

For a function $f(x)$ which has a discontinuous nth derivative this Fourier transform can be expressed as follows by integrating by parts n times, using the fact that any function $f(x)$ must equal 0 as $x \rightarrow \pm\infty$ for $\tilde{f}(k)$ to exist:

$$\widetilde{f^{(n)}}(k) = \frac{1}{\sqrt{2\pi}} \int_{-\infty}^{\infty} f^{(n)}(x)\, e^{ikx} dx \tag{5.8}$$

$$= \frac{1}{\sqrt{2\pi}} \left[\left[f^{(n-1)}(x) e^{ikx} \right]_{-\infty}^{\infty}{}^{\nearrow 0} - (ik) \int_{-\infty}^{\infty} f^{(n-1)}(x)\, e^{ikx} dx \right] \tag{5.9}$$

$$= \frac{-(ik)}{\sqrt{2\pi}} \left[\left[f^{(n-2)}(x) e^{ikx} \right]_{-\infty}^{\infty}{}^{\nearrow 0} - (ik) \int_{-\infty}^{\infty} f^{(n-2)}(x)\, e^{ikx} dx \right] \tag{5.10}$$

$$= \cdots = \pm (ik)^n\, \tilde{f}(k). \tag{5.11}$$

As $f^{(n)}(x)$ is discontinuous, we also know that:

$$\widetilde{f^{(n)}}(k) \sim \frac{1}{k}, \tag{5.12}$$

for large k, from the Fourier Transform of the Heaviside function. Therefore:

$$\widetilde{f^{(n)}}(k) \sim (ik)^n\, \tilde{f}(k) \sim \frac{1}{k} \tag{5.13}$$

$$\Rightarrow \tilde{f}(k) \sim \frac{1}{k^{n+1}}. \tag{5.14}$$

This means that the integral for kinetic energy in Fourier space will be:

$$k.e. \sim \int \frac{1}{k^{n+1}} \frac{k^2}{k^{n+1}} dk \tag{5.15}$$

$$\sim \int \left(\frac{1}{k^n} \right)^2 dk. \tag{5.16}$$

For good convergence of this integral the first derivative therefore needs to be continuous, and for even better convergence, which is desirable, the second derivative would also need to be continuous, i.e. in the previous derivation $n > 1$. This partially

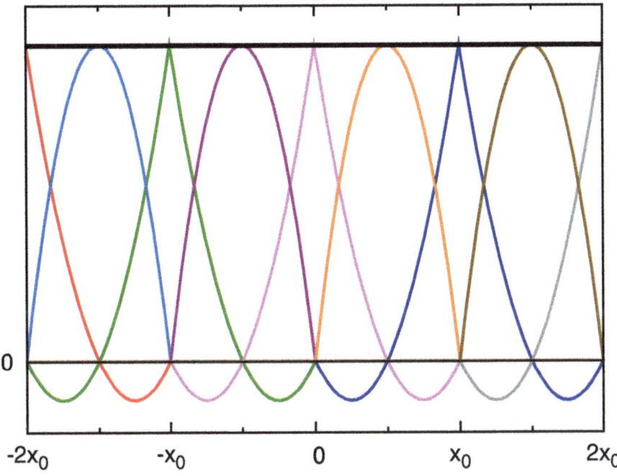

Fig. 5.2 Example of a one-dimensional quadratic finite element basis, with the basis elements shown in colour, and the sum over basis functions in *black*. As can be seen, all other basis functions are zero at the site of a cusp, and the extra functions further balance out the effect of the cusps so that the sum is constant

explains why the B-splines give such an accurate answer, as they are continuous up to the second derivative and thus rapid convergence would be expected.

There is one situation in which cusps might not affect the accuracy, when used in the finite element method [2]. However, in this case, more than one type of basis function is used, such that the total number of different types of functions is equal to the maximum degree of the polynomial. This means that, for a quadratic basis, for example, there will be two different basis functions, so that whenever any cusp arises all the other functions at that site are zero and hence the effect of the Dirac-δ function will be cancelled. See Fig. 5.2 for an example.

Shifting of the Eigenspectrum

Another effect worth studying is the discrepancy between the calculated and correct values for the lowest eigenvalue, into which insight can be gained by considering the coefficients of the eigenvector. The energy eigenvalue for the lowest band should be equal to zero, and all the coefficients c_0^α should be equal. Referring back to the definition of the wavefunction in terms of the coefficients in Eq. 5.1, we can write:

$$\sum_{\alpha=1}^{N} |c_0^\alpha|^2 = N|c_0|^2. \tag{5.17}$$

For the eigenfunction to be normalized, it then follows that:

$$N|c_0|^2 = 1 \tag{5.18}$$

$$\Rightarrow c_0 = \frac{1}{\sqrt{N}} = \sqrt{\frac{x_0}{L}}. \tag{5.19}$$

The corresponding eigenvalue can be found by taking the expectation value of the kinetic energy operator for the lowest eigenvalue $\mathcal{E}_0$:

$$\mathcal{E}_0 = \sum_{\alpha\beta} c_0^{\alpha} c_0^{\beta} T_{\alpha\beta} \tag{5.20}$$

$$= \sum_{\alpha\beta} T_{\alpha\beta} \frac{x_0}{L} \tag{5.21}$$

$$= \frac{x_0}{L} N \sum_{\beta} T_{\alpha\beta}, \tag{5.22}$$

as the sum over elements in a row will be the same for each row. The eigenvalue should be equal to zero, therefore this leads to the following condition on the kinetic energy matrix elements:

$$\sum_{\beta} T_{\alpha\beta} = 0. \tag{5.23}$$

No assumptions were made concerning the number of nearest neighbours, hence this condition applies equally to the nearest neighbour basis sets and the B-splines. As expected, it is obeyed by both the B-splines and the mirrored basis set, explaining why those band structures pass through zero as they should. Furthermore, for the orthogonal basis set this sum gives the exact energy value for the lowest band. For the other basis sets, however, there is still some additional shifting which needs further investigation.

It seems logical to consider next the mirrored basis set, as it fulfils this condition and so should be the easiest to understand. One consideration is how well the basis represents the eigenfunctions. It has been shown that the individual basis functions sum to zero, and it therefore seems likely that they can accurately represent a plane-wave, which can be verified by plotting the eigenfunctions as represented within the basis set.

Figure 5.3 shows the real part of an example eigenfunction for the mirrored basis. As shown in the magnified insert, the representation is not perfect, however as N is increased the curve becomes increasingly smooth, and thus will eventually become sufficiently smooth for any required level of accuracy. Therefore the problem is not with the eigenfunctions, which are guaranteed to respect the translational symmetry of the problem due to the manner in which the Hamiltonion is constructed. Rather it will be more useful to look more closely at the eigenvalues of the system as compared to the correct values. We first however observe that an inflection point can be seen between the points at which the basis functions are centred (in the insert of Fig. 5.3). This is due to the construction of symmetric nearest neighbour basis sets, which must necessarily have zero curvature at the point half way between neighbouring functions and zero slope at their centre points, unless one has introduced a cusp.

Referring back to Table 5.2, we can observe that the eigenvalues for the mirrored basis are approximately $\frac{4}{3}$ times the correct value, a relationship which becomes

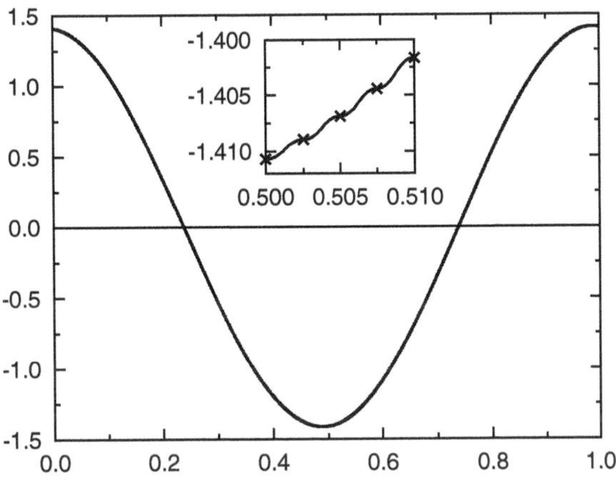

Fig. 5.3 The real part of the third lowest eigenfunction for the mirrored basis set for $N = 400$. The insert shows a magnified section, where the points are those at which a basis function is centred, and the line is the real space representation of the basis

Table 5.3 Matrix elements for the mirrored basis set before normalization

S_0	S_1	$T_0 x_0^2$	$T_1 x_0^2$	$\dfrac{P_1 x_0}{i}$
$\dfrac{23}{30}$	$\dfrac{7}{60}$	$\dfrac{4}{3}$	$-\dfrac{2}{3}$	$\dfrac{1}{2}$

increasingly exact as N is increased. The origins of this factor are not immediately obvious but a possible explanation becomes apparent when looking at the matrix elements before they are normalized, as shown in Table 5.3.

Looking at these values suggests that the factor of $\frac{4}{3}$ could be from the kinetic energy matrix elements. If the T_0 matrix element is arbitrarily changed to $1/x_0^2$, keeping the relationship with T_1 as defined from the condition in Eq. 5.23 and without changing any of the other matrix elements or re-normalizing, this does indeed correct the problems in the eigenspectrum, reproducing the expected free electron energies. Returning to the normalized matrix elements, it has been observed that this change to the matrix elements satisfies the following relationships:

$$T_0 x_0^2 = S_0 + 2S_1 \tag{5.24}$$

$$T_0 x_0 = 2\frac{P_1}{i}. \tag{5.25}$$

A derivation for the first of these relationships is given in Sect. 5.2.4. By arbitrarily defining sensible values for S_0 and S_1 and using these relationships to generate the rest of the matrix elements, a close approximation to the correct energy spectrum can always be achieved. However, they appear to be specific to nearest neighbour only basis sets, as B-splines do not follow the generalized relation $\sum_\beta S_{\alpha\beta} = T_0 x_0^2$. Furthermore, there seems to be no simple low order piecewise polynomial basis that satisfies the required conditions, implying that a nearest neighbour basis set is overly simplistic and unlikely to easily give accurate results.

In order to gain further insight into the reasons for the inaccurate eigenspectra generated with the different nearest neighbour basis sets, it is necessary to consider the eigenspectrum away from the Γ-point, and so we first turn to the calculation of band structures.

5.2 Generating Band Structures

We present here two methods for calculating band structures with localized basis functions. The presentation is limited to one dimension, for systems with local periodic potentials, however it can easily be extended to three dimensional systems with non-local potentials. Both are based on Bloch's theorem, the derivation for which was given in Sect. 2.3.2, however for convenience we reproduce both versions here in one dimension, where L is the length of the unit cell:

$$\psi_{nk}(x + L) = e^{ikL}\psi_{nk}(x) \tag{5.26}$$

and

$$\psi_{nk}(x) = e^{ikx}u_{nk}(x). \tag{5.27}$$

5.2.1 k · p Style Method

We call the first method for band structure calculation the $\mathbf{k} \cdot \mathbf{p}$ style method, as it relates to $\mathbf{k} \cdot \mathbf{p}$ perturbation theory. For this method we start from the second version of Bloch's theorem (5.27) and proceed by defining the periodic function $u_{nk}(x)$ as a sum over basis functions, $\{\phi_\alpha(x)\}$, with a set of k-dependent coefficients $\{c_{nk}^\alpha\}$:

$$u_{nk}(x) = c_{nk}^\alpha \phi_\alpha(x). \tag{5.28}$$

We are able to define a Hamiltonian which is dependent on k, starting from the original Schrödinger equation and proceeding as follows.

We first re-state the Schrödinger equation and substitute in our definition of the wavefunction in terms of $u_{ik}(x)$:

$$\hat{H}\psi_{nk}(x) = \mathcal{E}_{nk}\psi_{nk}(x) \tag{5.29}$$

$$\Rightarrow \hat{H}e^{ikx}u_{nk}(x) = \mathcal{E}_{nk}e^{ikx}u_{nk}(x). \tag{5.30}$$

Premultiplying by e^{-ikx}, we can then define a k-dependent Hamiltonian:

$$e^{-ikx}\hat{H}e^{ikx}u_{nk}(x) = \mathcal{E}_{nk}u_{nk}(x) \tag{5.31}$$

$$\Rightarrow \hat{H}(k)u_{nk}(x) = \mathcal{E}_{nk}u_{nk}(x). \tag{5.32}$$

In order to work out the correct form of this new Hamiltonian, we first insert our definition of $u_{nk}(x)$ in terms of the basis functions and re-write the eigenvalue equation in integral form:

$$\int_0^L \phi_\alpha^*(x)\,e^{-ikx}\,\hat{H}\,e^{ikx}\,c_{nk}^\beta\,\phi_\beta(x)\,dx = \mathcal{E}_{nk}\int_0^L \phi_\alpha^*(x)\,c_{nk}^\beta\,\phi_\beta(x)\,dx \tag{5.33}$$

$$\Rightarrow H_{\alpha\beta}(k)\,c_{nk}^\beta = \mathcal{E}_{nk}\,S_{\alpha\beta}c_{nk}^\beta. \tag{5.34}$$

By writing the Hamiltonian operator as a sum of the kinetic and potential energy operators, we can thus derive a form for the k-dependent Hamiltonian:

$$H_{\alpha\beta}(k) = \int_0^L \phi_\alpha^*(x)\,e^{-ikx}\left(\hat{T} + \hat{V}\right)e^{ikx}\phi_\beta(x)\,dx \tag{5.35}$$

$$= \int_0^L \phi_\alpha^*(x)\,e^{-ikx}\left(-\frac{1}{2}\frac{d^2}{dx^2} + V(x)\right)e^{ikx}\phi_\beta(x)\,dx \tag{5.36}$$

$$= \frac{1}{2}k^2\int_0^L \phi_\alpha^*(x)\,\phi_\beta(x)\,dx - ik\int_0^L \phi_\alpha^*(x)\,\frac{d\phi_\beta}{dx}dx \tag{5.37}$$

$$- \frac{1}{2}\int_0^L \phi_\alpha^*(x)\,\frac{d^2\phi_\beta}{dx^2}dx + \int_0^L \phi_\alpha^*(x)\,V(x)\,\phi_\beta(x)\,dx. \tag{5.38}$$

Returning to the matrix element definitions in Eqs. 5.4, 5.5 and 5.6, and introducing an equivalent definition for the potential matrix elements, which we shall denote $V_{\alpha\beta}$, we have:

$$H_{\alpha\beta}(k) = \frac{1}{2}k^2 S_{\alpha\beta} - k P_{\alpha\beta} + T_{\alpha\beta} + V_{\alpha\beta}. \tag{5.39}$$

Therefore, in order to generate a band structure, Eq. 5.34 should be solved for a number of different k values, generating the Hamiltonian according to Eq. 5.39. For the nearest neighbour basis sets, the Hamiltonian matrix will have the following tridiagonal structure:

$$\begin{pmatrix} H_0 & H_1 & 0 & 0 & H_1^* \\ H_1^* & H_0 & H_1 & 0 & 0 \\ 0 & H_1^* & H_0 & H_1 & 0 \\ 0 & 0 & H_1^* & H_0 & H_1 \\ H_1 & 0 & 0 & H_1^* & H_0 \end{pmatrix}, \tag{5.40}$$

where the corner elements are present to enforce periodic boundary conditions and the basis functions are considered to be fully periodic, as demonstrated by Fig. 5.4a.

k · p Perturbation Theory

The name of this method derives from $\mathbf{k}\cdot\mathbf{p}$ perturbation theory, where the dependence of the Hamiltonian on k is treated as a perturbation due to taking a small step in reciprocal space δk. Again following Bloch's theorem, the Hamiltonian is defined (in one dimension) as:

$$\hat{H}(k) = \mathrm{e}^{-\mathrm{i}kx}\hat{H}\mathrm{e}^{\mathrm{i}kx}. \tag{5.41}$$

The perturbation to the Hamiltonian can be found via a Taylor expansion of the exponentials in Eq. 5.41 such that:

$$\Delta\hat{H}(k) = \hat{H}(k+\delta k) - \hat{H}(k) \tag{5.42}$$

$$= \mathrm{e}^{-\mathrm{i}kx}\left(\mathrm{i}\delta k\left[\hat{H},x\right] - \frac{1}{2}(\delta k)^2\left[\left[\hat{H},x\right],x\right] + O\left(\delta k^3\right)\right)\mathrm{e}^{\mathrm{i}kx}$$

and likewise for the overlap operator $\hat{S}$, should one be present. This can then be compared with the general expressions for first and second order changes in energy given a perturbation in both $\hat{H}$ and $\hat{S}$, for example:

$$\mathcal{E}_n^{(1)} = \langle\psi_n^{(0)}|\Delta\hat{H}^{(1)} - \mathcal{E}_n^{(0)}\Delta\hat{S}^{(1)}|\psi_n^{(0)}\rangle, \tag{5.43}$$

where $\left\{|\psi_n^{(0)}\rangle\right\}$ are the eigenfunctions of the original, unperturbed system.

The first and second order expressions for energy can then be used to find information about the shape of a given band structure around the Γ-point without having to solve the matrix equation at a number of different k values, making it a useful method when detailed knowledge of the entire band structure is unnecessary. For a more detailed derivation and an example of its usefulness see [3, 4].

This differs from our method in that $\mathbf{k}\cdot\mathbf{p}$ perturbation theory uses the information at a single $\mathbf{k}$-point to calculate the eigenvalues at a neighbouring point. Our $\mathbf{k}\cdot\mathbf{p}$ style method, on the other hand, uses the same principle to construct a new Hamiltonian at each $\mathbf{k}$-point, and thus calculate the eigenvalues without resorting to the use of perturbation theory.

5.2.2 Tight-Binding Style Method

The second method is based on the first statement of Bloch's theorem (Eq. 5.26) and results in an expression for the wavefunction which relates to that used in tight-binding. Unlike the $\mathbf{k}\cdot\mathbf{p}$ style method, the basis functions are not considered to be fully periodic functions with a repeat length of the unit cell, rather each basis function is separate and fully localized, as demonstrated by Fig. 5.4b. Following Bloch's theorem, we can form a set of k-dependent extended basis functions $\left\{\phi'_{\alpha k}(x)\right\}$ by summing over the original basis functions $\{\phi_\alpha(x)\}$:

$$\phi'_{\alpha k}(x) = \sum_m \phi_\alpha(x - mL)\,e^{ikmL}. \tag{5.44}$$

The wavefunction can then be written as an appropriately weighted sum over these new basis functions in the following manner:

$$\psi_{nk}(x) = c^\alpha_{nk}\phi'_{\alpha k}(x) \tag{5.45}$$

$$= \sum_m c^\alpha_{nk}\phi_\alpha(x - mL)\,e^{ikmL}, \tag{5.46}$$

which, by construction, must agree with Bloch's theorem. We can also relate this to the second version of Bloch's theorem by writing:

$$\psi_{nk}(x) = e^{ikx}e^{-ikx}\sum_m c^\alpha_{nk}\phi_\alpha(x - mL)\,e^{ikmL} \tag{5.47}$$

$$= e^{ikx}\sum_m c^\alpha_{nk}\phi_\alpha(x - mL)\,e^{-ik(x-mL)}, \tag{5.48}$$

so that the periodic function $u_{nk}(x)$ is:

$$u_{nk}(x) = \sum_m c^\alpha_{nk}\phi_\alpha(x - mL)\,e^{-ik(x-mL)}. \tag{5.49}$$

Using these definitions, we can determine the form of the Hamiltonian and overlap matrix elements, starting again from the appropriate Schrödinger equation:

$$\hat{H}\psi_{nk}(x) = \mathcal{E}_{nk}\psi_{nk}(x) \tag{5.50}$$

$$\Rightarrow \hat{H}\sum_m c^\alpha_{nk}\phi_\alpha(x - mL)\,e^{ikmL} = \mathcal{E}_{nk}\sum_m c^\alpha_{nk}\phi_\alpha(x - mL)\,e^{ikmL}. \tag{5.51}$$

Premultiplying by the extended k-dependent basis function and integrating over the unit cell, we then have:

$$\int_0^L \sum_p \phi^*_\alpha(x - pL)\,e^{-ikpL}\,\hat{H}\sum_m c^\beta_{nk}\phi_\beta(x - mL)\,e^{ikmL}\,dx$$

$$= \mathcal{E}_{nk}\int_0^L \sum_p \phi^*_\alpha(x - pL)\,e^{-ikpL}\sum_m c^\beta_{nk}\phi_\beta(x - mL)\,e^{ikmL}\,dx \tag{5.52}$$

$$\Rightarrow \sum_{pm} e^{-ikL(p-m)}c^\beta_{nk}\int_0^L \phi^*_\alpha(x - pL)\,\hat{H}\phi_\beta(x - mL)\,dx$$

$$= \mathcal{E}_{nk}\sum_{pm} e^{-ikL(p-m)}c^\beta_{nk}\int_0^L \phi^*_\alpha(x - pL)\,\phi_\beta(x - mL)\,dx. \tag{5.53}$$

As the basis functions are highly localized, only those basis functions right at the edge of the unit cell will overlap with those in other unit cells, so except for these matrix elements, when $p \neq m$ both $H_{\alpha\beta}$ and $S_{\alpha\beta}$ will equal zero. For the nearest neighbour basis sets, the only exception will be between $\phi_1(x)$ and $\phi_N(x + L)$, which will be calculated via the relation:

$$H_{N1} = \sum_{pm} e^{-ikL(p-m)} \int_0^L \phi_N^*(x - pL)\,\hat{H}\phi_1(x - mL)\,\mathrm{d}x \tag{5.54}$$

$$= e^{ikL} \int_0^L \phi_N(x + L)\,\hat{H}\phi_1(x)\,\mathrm{d}x \tag{5.55}$$

and similarly for the overlap matrix. As with the $\mathbf{k} \cdot \mathbf{p}$ style method, the matrices $H_{\alpha\beta}$ and $S_{\alpha\beta}$ are constructed for different k-points and the generalized eigenvalue equation is then solved to calculate a band structure.

5.2.3 Free Electron Band Structures

Having previously calculated the eigenvalues at the Γ-point only, band structures were now calculated for all the basis sets using both methods. The aim was both to compare the two methods of band structure calculation, and to further understand the criteria for a good localized basis set. The band structures for the orthogonal, cusp-free and mirrored basis sets are compared to the correct zero potential band structure in Fig. 5.5, using $N = 400$ basis functions in a unit cell of size $L = 1$ Bohr. The B-spline results were indistinguishable from the correct results and the orthogonal splines and extra cusps results showed similar disagreements with the true band structure to the other nearest neighbour basis sets.

Despite the clear differences from the correct band structure for the nearest neighbour basis sets, some conclusions can be drawn about the differences between the methods for band structure calculation. Notably, the $\mathbf{k} \cdot \mathbf{p}$ style method gives a more accurate representation of the shape of the bands, whereas the tight-binding style method forces the bands to meet at the Brillouin zone boundary due to the inherent periodicity of the method. Therefore, for incomplete basis sets the $\mathbf{k} \cdot \mathbf{p}$ style method appears to be more applicable when the shape of a given band is required to be accurate, whereas the tight-binding style method would be better when the study of band gaps is more important.

5.2.4 Further Requirements for a Good Basis Set

Following on from earlier analysis in Sect. 5.1.2, the discrepancies between the calculated band structures in the nearest neighbour basis sets and the correct result allow

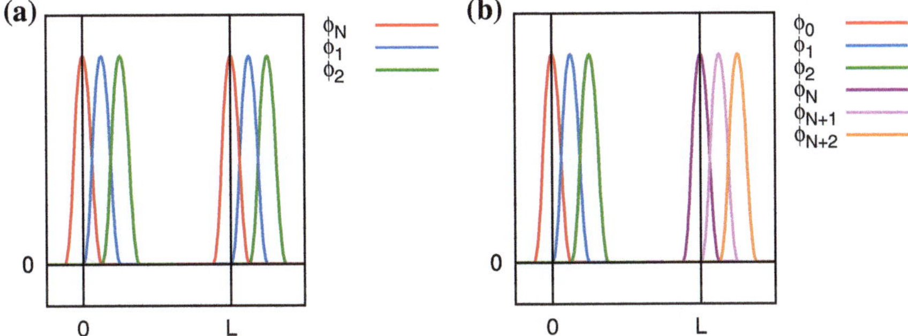

Fig. 5.4 Schematics demonstrating the difference in the definition of the basis functions in the two methods for band structure calculation. For the $\mathbf{k} \cdot \mathbf{p}$ style method the functions centred at $x = 0$ and $x = L$ are a single function, whereas with the tight-binding style method they are two separate functions. Therefore for the tight-binding style method when calculating overlap and Hamiltonian matrix elements within the unit cell between for example ϕ_1 and ϕ_N, a phase factor must be included to translate ϕ_N to $x = 0$

one to gain further insight into the requirements for a good basis set. Given that the translational symmetry of the problem leads to the solutions being plane-waves of the form $e^{iq_m x_0}$ where $qL = 2\pi n$, one can analytically solve the eigenvalue problem for both methods for a nearest neighbour basis set.

$\mathbf{k} \cdot \mathbf{p}$ Style Method

For the $\mathbf{k} \cdot \mathbf{p}$ method in an orthogonal basis this gives the following expression for the energy of a given band $\mathcal{E}_{nk}$:

$$\mathcal{E}_{nk} = \frac{1}{2}k^2 + T_0 + 2T_1 \cos\left(\frac{2\pi n}{N}\right) + 2P_1 \sin\left(\frac{2\pi n}{N}\right). \tag{5.56}$$

Requiring the lowest band $n = 0$ to have zero energy at the Γ-point, we again recover the condition $T_0 + 2T_1 = 0$. This also highlights the fact that there is no inherent periodicity in the band structure calculated using this method, as there is no dependence on the length of the unit cell and therefore no periodicity equivalent to the Brillouin zone.

Repeating this simple analysis for a non-orthogonal basis gives:

$$\mathcal{E}_{nk} = \frac{\frac{1}{2}k^2\left[S_0 + 2S_1\cos\left(\frac{2\pi n}{N}\right)\right] + T_0 + 2T_1\cos\left(\frac{2\pi n}{N}\right) + 2P_1\sin\left(\frac{2\pi n}{N}\right)k}{S_0 + 2S_1\cos\left(\frac{2\pi n}{N}\right)}, \tag{5.57}$$

which gives no new information when considering the lowest band, assuming as before that $T_0 + 2T_1 = 0$.

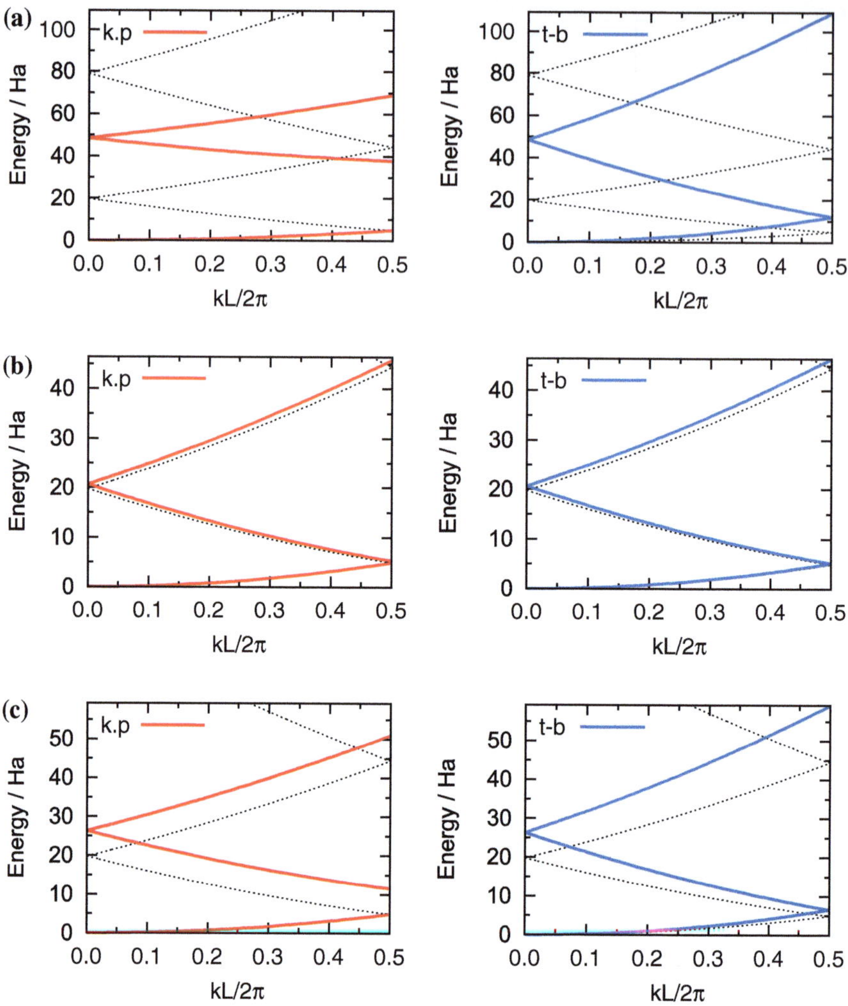

Fig. 5.5 Band structures for a zero potential system for three nearest neighbour basis sets calculated using both the $\mathbf{k}\cdot\mathbf{p}$ and tight-binding (t-b) methods, with $N = 400$. The basis set results are translated vertically by the values given in Table 5.2 for comparison with the correct result, which is indicated by the *dotted lines*

Tight-Binding Style Method

An equivalent analysis for the tight-binding style method gives the following energy expression:

$$\mathcal{E}_{nk} = T_0 + 2T_1 \cos\left(kx_0 + \frac{2\pi n}{N}\right). \tag{5.58}$$

This clearly has a periodicity in the Brillouin zone which is dependent on the width of the basis functions x_0 and therefore also the length of the unit cell, as $L = Nx_0$. However, to demonstrate free electron dispersion of the lowest band around the Γ-point, the following approximation to this expression can be made:

$$\mathcal{E}_{0k} = T_0 + 2T_1\cos(kx_0) \tag{5.59}$$

$$= (T_0 + 2T_1) - 2T_1(1 - \cos(kx_0)) \tag{5.60}$$

$$\simeq (T_0 + 2T_1) - T_1 x_0^2 k^2, \tag{5.61}$$

therefore implying that:

$$T_1 x_0^2 = -\frac{1}{2}. \tag{5.62}$$

For a non-orthogonal basis we have:

$$\mathcal{E}_{nk} = \frac{T_0 + 2T_1\cos\left(kx_0 + \frac{2\pi n}{N}\right)}{S_0 + 2S_1\cos\left(kx_0 + \frac{2\pi n}{N}\right)}, \tag{5.63}$$

which, following the same simplification, then Taylor expanding the denominator up to second order in k gives:

$$\mathcal{E}_{0k} \simeq \frac{(T_0 + 2T_1) - T_1 x_0^2 k^2}{(S_0 + 2S_1) - S_1 x_0^2 k^2} \tag{5.64}$$

$$\simeq \frac{(T_0 + 2T_1)}{(S_0 + 2S_1)}\left(1 - \frac{T_1}{T_0 + 2T_1}x_0^2 k^2\right)\left(1 + \frac{S_1}{S_0 + 2S_1}x_0^2 k^2\right) \tag{5.65}$$

$$\simeq \frac{T_0 + 2T_1}{S_0 + 2S_1} - x_0^2 k^2\left[\frac{T_1}{S_0 + 2S_1} - \frac{S_1(T_0 + 2T_1)}{(S_0 + 2S_1)^2}\right] \tag{5.66}$$

$$\simeq -x_0^2 k^2 \frac{T_1}{S_0 + 2S_1}, \tag{5.67}$$

once again using the previous condition of $T_0 + 2T_1 = 0$. Therefore this gives $S_0 + 2S_1 = -2T_1 x_0^2$, so that for example in the orthogonal case $-2T_1 x_0^2 = 1 \Rightarrow T_1 x_0^2 = -\frac{1}{2}$, as already found. This is consistent with the original relationships, although it implies that the fundamental relationship is between T_1 and the overlap matrix elements, not T_0 as originally suggested.

Verification of the Analysis

These expressions were tested by comparing the numerical and analytical results for the mirrored basis and the cusp-free basis. Excellent agreement was obtained between the predicted band structures and those obtained by solving the eigenvalue problem using direct diagonalization. The new conditions derived were used to define a set of matrix elements for an ideal basis, with $S_0 = \frac{3}{5}$, $S_1 = \frac{1}{5}$, $P_1 = \frac{i}{2x_0}$, $T_0 = \frac{1}{x_0^2}$ and $T_1 = -\frac{1}{2x_0^2}$, which gave the correct band structure as expected, however these matrix elements do not necessarily correspond to a real basis set.

These relationships can be generalized to a non-nearest neighbour basis set using the same procedure as above for the tight-binding based method, which will give the following for $\mathcal{E}_{0k}$:

$$\mathcal{E}_{0k} = \frac{T_0 + 2T_1\cos(kx_0) + 2T_1\cos(2kx_0) + 2T_3\cos(3kx_0) + \cdots}{S_0 + 2S_1\cos(kx_0) + 2S_2\cos(2kx_0) + 2S_3\cos(3kx_0) + \cdots} \tag{5.68}$$

$$\simeq \frac{\sum_\beta T_{\alpha\beta} - \frac{1}{2}x_0^2 k^2 \sum_\beta \beta^2 T_{\alpha\beta}}{\sum_\beta S_{\alpha\beta} - \frac{1}{2}x_0^2 k^2 \sum_\beta \beta^2 S_{\alpha\beta}} \tag{5.69}$$

$$\simeq \frac{-\frac{1}{2}x_0^2 k^2 \sum_\beta \beta^2 T_{\alpha\beta}}{\sum_\beta S_{\alpha\beta}} \tag{5.70}$$

$$\Rightarrow \sum_\beta S_{\alpha\beta} = -x_0^2 \sum_\beta \beta^2 T_{\alpha\beta}. \tag{5.71}$$

This relationship holds true for the B-splines, and also gives the correct band structure when used to arbitrarily define matrix elements within the tight-binding based method. However, some generalization is still needed to find a condition on the momentum matrix elements (or prove Eq. 5.25), which can probably only be achieved by considering more than just the lowest band for the $\mathbf{k} \cdot \mathbf{p}$ style method, for which the analysis would be less straightforward. Furthermore, knowledge of the required relationships between matrix elements does not guarantee the existence of such a basis, therefore it would seem likely that B-splines are one of the simplest low-order piecewise polynomial localized basis sets that are capable of accurately reproducing the free-electron band structure. This will be verified by considering systems with simple potentials in the following Section.

5.2.5 Simple Potentials

Two simple one-dimensional potentials were implemented in the test program; Kronig-Penney [5] and Gaussian potentials. The Kronig-Penney potential is defined as:

$$V(x) = \begin{cases} V_0 & \text{if } 0 < x < b \\ 0 & \text{if } b < x < b + w \end{cases}$$

with a width w, height V_0 and centred at $b + w/2$, so that the length of the unit cell is $L = b + w$. The solution can be derived analytically [5], which provides a useful basis for comparison. The Gaussian potential is defined as:

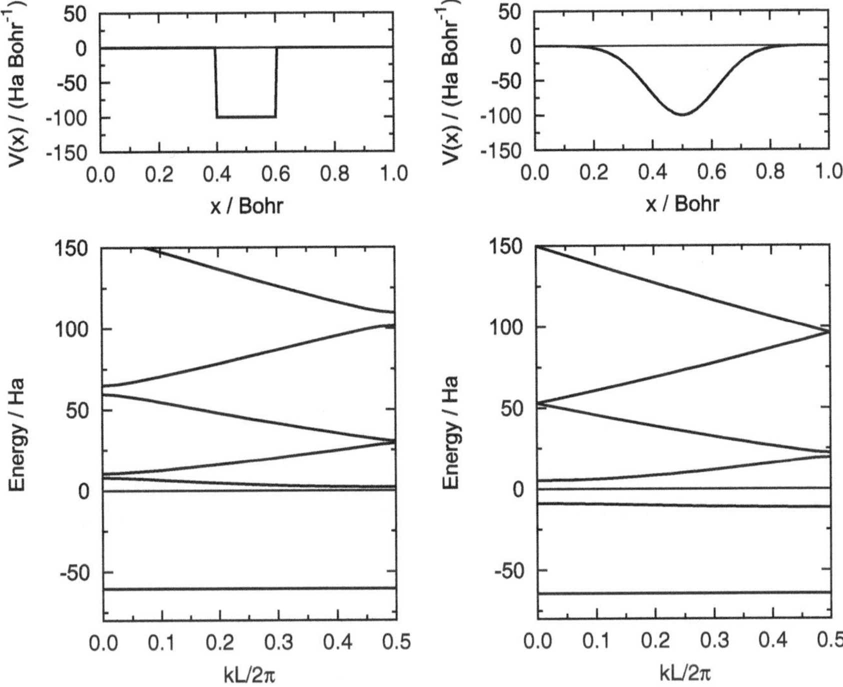

Fig. 5.6 Example Kronig-Penney [*left*] and Gaussian [*right*] potentials and band structures, calculated with $N = 200$ in the B-spline basis

$$V(x) = V_0 e^{-\frac{(x-b)^2}{2c^2}} \tag{5.72}$$

where V_0 is the height, b defines the centre and c is related to the full width at half maximum in the usual manner. Examples of both types of potential are shown in Fig. 5.6, with the corresponding band structures, as calculated in the B-spline basis.

In Fig. 5.7, a small, unconverged set of B-splines is used to calculate the band structure using both methods for the two different types of potentials. The lack of convergence highlights the difference between the methods, where the tight-binding method gives a better approximation of the band gap and the $\mathbf{k} \cdot \mathbf{p}$ method gives a better approximation of the band shape. Whilst this behaviour agrees with earlier observations made for "bad" basis sets in a zero potential system, in this case it is much more approximate, with neither the correct gap nor the correct shape achieved with either of the two methods. This highlights the need for good basis set convergence when calculating band structures. However, we do note that the tight-binding style method gives a more faithful representation of the true result, and would therefore probably be the better method to use in ONETEP.

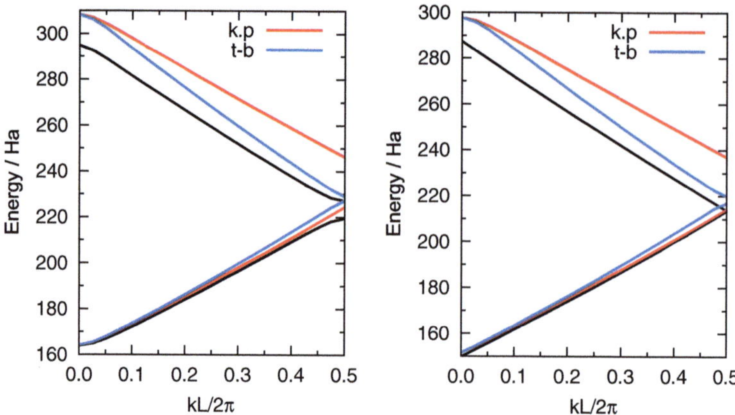

Fig. 5.7 The 7th and 8th bands for Kronig-Penney [*left*] and Gaussian [*right*] potentials using an unconverged basis set with $N = 10$ B-spline functions, compared to the converged result, which is depicted in black. Both the $\mathbf{k} \cdot \mathbf{p}$ and tight-binding (t-b) methods are compared

5.3 Summary

We have studied a number of nearest neighbour localized basis sets, including the consideration of necessary conditions on a "good" basis set. In particular, it was determined that the existence of cusps in the localized basis sets, which are similar to the kinks that can occur in NGWFs in ONETEP at the edge of their localization regions, has a detrimental effect on the accuracy of resulting band structures. Some conditions have been derived for the matrix elements of a good basis set, however no nearest neighbour basis set was found which satisfied all of these conditions, and it seems likely that a nearest neighbour formulation is overly simplistic. Instead, B-splines, which are localized to third nearest neighbours, were demonstrated to accurately reproduce band structures for all the systems studied and so these have been adopted for the work in the following chapter.

Two different methods for generating band structures using a localized basis set have been investigated, a $\mathbf{k} \cdot \mathbf{p}$ perturbation theory based method and a tight-binding based method. For a good basis set, which is fully converged, the two methods agree. However, for an imperfect basis set or for an insufficient number of basis functions, there are significant differences between the two methods. The $\mathbf{k} \cdot \mathbf{p}$ method gives a better approximation of the shape of the bands, however it can produce highly inaccurate band gaps. The tight-binding based method, on the other hand, better enforces the correct periodicity and so achieves more accurate results for gaps between bands. Whilst the calculated band shapes are less accurate than with the $\mathbf{k} \cdot \mathbf{p}$ style method, on average the tight-binding style method gives better results.

References

1. Hernández, E., Gillan, M.J., Goringe, C.M.: Phys. Rev. B **55**(20), 13485 (1997)
2. Pask, J.E., Sterne, P.A.: Modell. Simul. Mater. Sci. Eng. **13**(3), R71 (2005)
3. Pickard, C.J., Payne, M.C.: Phys. Rev. B **62**(7), 4383 (2000)
4. Ashcroft, N.W., Mermin, N.D.: Solid State Physics. Holt, Rinehart and Winston (1976)
5. C. Kittel: Introduction to Solid State Physics, 7th edn. Wiley, New York (1996)

Chapter 6
Conduction States: Methods and Applications

As discussed in Sect. 3.4.5, the unoccupied Kohn-Sham states tend not to be well represented within the basis of NGWFs which are optimized during a ground state calculation. Therefore it is necessary to optimize a second set of NGWFs to represent the unoccupied states in a non self-consistent calculation. In this Chapter we present three different methods for doing so and identify the projection method as the most promising. We go on to describe the implementation of the projection method in ONETEP, including the description of a scheme designed to avoid problems associated with local minima. We also describe the application of the method to the calculation of optical absorption spectra in ONETEP.

Parts of this Chapter have been previously published in the following paper: L. E. Ratcliff, N. D. M. Hine and P. D. Haynes, **Calculating optical absorption spectra for large systems using linear-scaling density-functional theory**, *Phys. Rev. B* **84**, 165131 (2011), Copyright (2011) by the American Physical Society [1].

6.1 The Different Methods

Possible methods for optimizing a new set of NGWFs to represent the conduction states include the folded spectrum method [2, 3], the shift-invert method [4] and the use of a projection operator. These differ principally by the form of the eigenvalue equation they attempt to solve to obtain the excited states. These were implemented in the toy model described in the previous Chapter and both the accuracy and efficiency of each method was compared.

6.1.1 Folded Spectrum

The folded spectrum method involves folding the energy spectrum of a matrix $\mathbf{H}$ around a reference energy E_{ref}, where the spectrum of $\mathbf{H}$ is found from the eigenvalue

L. Ratcliff, *Optical Absorption Spectra Calculated Using Linear-Scaling Density-Functional Theory*, Springer Theses, DOI: 10.1007/978-3-319-00339-9_6,
© Springer International Publishing Switzerland 2013

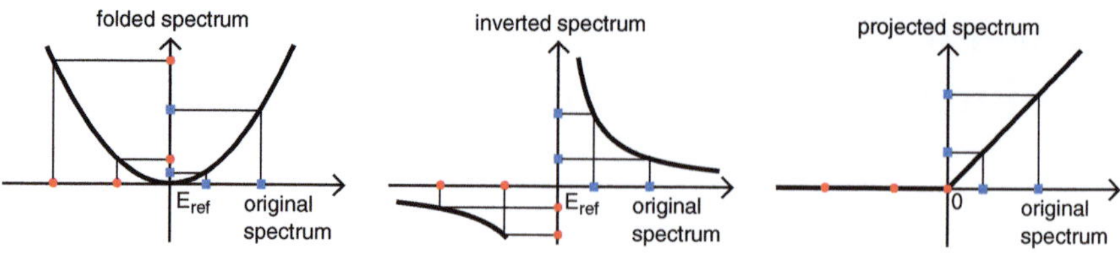

Fig. 6.1 Schematics comparing the three methods for the calculation of unoccupied states. The original spectrum is shown on the x-axis and the transformed spectrum on the y-axis, with *thick black curves* depicting the relationship between the two sets of eigenvalues. The occupied states are shown as *red circles* and the unoccupied states as *blue squares*, with the reference energy arbitrarily chosen to be in the gap for the folded spectrum and shift invert methods

equation $\mathbf{H}\mathbf{x} = \mathcal{E}\mathbf{x}$. This leads to a new eigenvalue equation which the eigensolutions of the original equation also satisfy:

$$(\mathbf{H} - E_{\text{ref}}\mathbf{I})^2\,\mathbf{x} = (\mathcal{E} - E_{\text{ref}})^2\,\mathbf{x}. \tag{6.1}$$

The smallest eigenvalues of this new matrix are related to those of $\mathbf{H}$ nearest E_{ref}, so that by setting E_{ref} to a value near the centre of the energy range covered by the conduction states, they can be found by solving the new eigenvalue equation.

The transformed equation could be used to find the eigenvalues of the original equation directly, as the eigenvalues of the new equation $\mathcal{E}'$ are defined in relation to $\mathcal{E}$, such that $\mathcal{E}' = (\mathcal{E} - E_{\text{ref}})^2$. However in practice, one instead calculates $\mathbf{x}^*\mathbf{H}\mathbf{x}$, to avoid the need for knowing whether or not the eigenvalues found correspond to those above or below the reference energy in the original problem, i.e. whether to calculate $\mathcal{E} = E_{\text{ref}} + \sqrt{\mathcal{E}'}$ (for the case where $\mathcal{E} > E_{\text{ref}}$) or $\mathcal{E} = E_{\text{ref}} - \sqrt{\mathcal{E}'}$ (for the case where $\mathcal{E} < E_{\text{ref}}$).

The folded spectrum method can also be generalized to account for the use of a non-orthogonal basis set [5], giving the following:

$$(\mathbf{H} - E_{\text{ref}}\mathbf{S})\,\mathbf{S}^{-1}\,(\mathbf{H} - E_{\text{ref}}\mathbf{S})\,\mathbf{x} = (\mathcal{E} - E_{\text{ref}})^2\,\mathbf{S}\mathbf{x}. \tag{6.2}$$

This is illustrated by Fig. 6.1, which contains a schematic showing the effect of the folded spectrum method on a set of example eigenvalues. This method has been used previously for example to study the conduction band minimum for Si within the tight-binding method [6], as well as in studies of quantum dots [7, 8].

6.1.2 Shift Invert

Shift invert is another method of spectral transformation which can be used to find extremal eigenvalues. It can also be used to convert a generalized eigenvalue equation to a standard eigenvalue equation. Starting from the generalized eigenvalue equation

$\mathbf{Hx} = \mathcal{E}\mathbf{Sx}$, the following simple transformation can be made:

$$\mathbf{H}^{-1}\mathbf{Hx} = \mathcal{E}\mathbf{H}^{-1}\mathbf{Sx} \tag{6.3}$$

$$\Rightarrow \mathbf{x} = \mathcal{E}\mathbf{H}^{-1}\mathbf{Sx} \tag{6.4}$$

$$\Rightarrow \mathbf{H}^{-1}\mathbf{Sx} = \frac{1}{\mathcal{E}}\mathbf{x}. \tag{6.5}$$

Adding a reference energy in a similar manner to the folded spectrum method thus gives the generalized shift invert equation:

$$(\mathbf{H} - E_{\mathrm{ref}}\mathbf{S})^{-1}\mathbf{Sx} = \frac{1}{\mathcal{E} - E_{\mathrm{ref}}}\mathbf{x}. \tag{6.6}$$

As can be seen, this shift invert equation is now a standard, rather than a generalized eigenvalue equation, which can be a very useful transformation. However, even if both $\mathbf{S}$ and $\mathbf{H}$ are Hermitian, $(\mathbf{H} - E_{\mathrm{ref}}\mathbf{S})^{-1}\mathbf{S}$ will not generally be Hermitian [9, 10], which could result in decreased numerical efficiency when solving the equation. The most straightforward method of ensuring that the transformed Hamiltonian is Hermitian is to pre-multiply by the overlap matrix, giving:

$$\mathbf{S}\,(\mathbf{H} - E_{\mathrm{ref}}\mathbf{S})^{-1}\mathbf{Sx} = (\mathcal{E} - E_{\mathrm{ref}})^{-1}\mathbf{Sx}. \tag{6.7}$$

For this case, the eigenvalues of the original matrix will be calculated in descending order, starting from the reference energy, as demonstrated in Fig. 6.1, which contains a diagram showing the transformation of a set of example eigenvalues following the application of shift invert. In order to correctly calculate the conduction states, the reference energy should therefore be set between the highest required conduction band and the state immediately above (shift invert variant +). One way to avoid this problem is to multiply the new Hamiltonian by minus one, reversing the order of calculation and therefore allowing the conduction states to be calculated in ascending order starting from the LUMO, simply by setting the reference energy to be just above the HOMO energy (shift invert variant $-$).

The shift invert method can suffer from stability problems, which can be reduced by adding an imaginary component, $i\mu$, to the reference energy, however this means that the Hamiltonian once again loses its Hermiticity, creating the possibility of complex eigenvalues. This can be avoided by combining two shift invert transformations, such that a small positive imaginary component is added to the reference energy for the first transformation and a negative component is added to the second, thereby eliminating all imaginary components:

$$[\mathbf{H} - (E_{\mathrm{ref}} - \mu i)\,\mathbf{S}]\,\mathbf{x} = (\mathcal{E} - (E_{\mathrm{ref}} - \mu i))\,\mathbf{Sx} \tag{6.8}$$

$$\Rightarrow [\mathbf{H} - (E_{\mathrm{ref}} + \mu i)\,\mathbf{S}]\,\mathbf{S}^{-1}\,[\mathbf{H} - (E_{\mathrm{ref}} - \mu i)\,\mathbf{S}]\,\mathbf{x} = (\mathcal{E} - (E_{\mathrm{ref}} + \mu i))$$
$$\times (\mathcal{E} - (E_{\mathrm{ref}} - \mu i))\,\mathbf{Sx} \tag{6.9}$$

$$\left[\mathbf{HS}^{-1}\mathbf{H} - \mathbf{HS}^{-1}\left(E_{\text{ref}} + \mu i\right)\mathbf{S} - \left(E_{\text{ref}} - \mu i\right)\mathbf{SS}^{-1}\mathbf{H}\right.$$
$$\left. - \left(E_{\text{ref}} - \mu i\right)\mathbf{SS}^{-1}\left(E_{\text{ref}} + \mu i\right)\mathbf{S}\right]\mathbf{x} = \left(\mathcal{E}^2 - 2E_{\text{ref}}\mathcal{E}\right.$$
$$\left. + E_{\text{ref}}^2 + \mu^2\right)\mathbf{Sx} \quad (6.10)$$
$$\left[\mathbf{HS}^{-1}\mathbf{H} - 2E_{\text{ref}}\mathbf{H} + \left(E_{\text{ref}}^2 + \mu^2\right)\mathbf{S}\right]\mathbf{x} = \left(\mathcal{E}^2 - 2E_{\text{ref}}\mathcal{E}\right.$$
$$\left. + E_{\text{ref}}^2 + \mu^2\right)\mathbf{Sx}. \quad (6.11)$$

This can be rearranged to give the final generalized eigenvalue equation:

$$\mathbf{S}\left[\mathbf{HS}^{-1}\mathbf{H} - 2E_{\text{ref}}\mathbf{H} + \left(E_{\text{ref}}^2 + \mu^2\right)\mathbf{S}\right]^{-1}\mathbf{Sx} = \left(\mathcal{E} - E_{\text{ref}}\right)^{-2}\mathbf{x}, \quad (6.12)$$

where we have neglected the μ^2 term from the eigenvalue denominator, as μ is very small. In this case the eigenvalues appear in an unfavourable order, such that as the transformed eigenvalues increase in energy, $|\mathcal{E} - E_{\text{ref}}|$ decreases, i.e. the eigenvalues furthest from E_{ref} will be found first. Multiplying the Hamiltonian by minus one will reverse the order, returning to the situation where eigenvalues closest to the reference energy are found first (shift invert variant i). This resembles the folded spectrum method in that the conduction and valence states again become mixed, and so a careful choice of reference energy is needed.

6.1.3 Projection

The density operator is defined according to Eq. 3.3, where we assume the f_n to be 1 for valence states and 0 for conduction states within the test program. The density operator $\hat{\rho}$ is a projection operator onto the subspace of states occupied by the valence states, so that projecting $\hat{\rho}$ onto $\hat{H}$ and solving the new eigenvalue equation will give only the valence eigenstates. Alternatively, projecting with $\hat{I} - \hat{\rho}$, where for ONETEP the $\hat{I}$ is defined in the underlying psinc basis, will leave only contributions from the conduction states. This is illustrated in Fig. 6.1, which contains a schematic demonstrating the effect of projecting the Hamiltonian in this manner on a set of example eigenvalues. In this way, a new eigenvalue equation could be solved to find the correct conduction states, which for ONETEP will correspond to minimizing a set of conduction NGWFs with respect to a total energy expression involving the projected Hamiltonian.

One problem which can arise due to the imposition of localization constraints during a calculation is that $\hat{H}$ and $\hat{\rho}$ may not commute exactly, which will result in the projected Hamiltonian no longer being Hermitian. This can be overcome by projecting twice, so that the expression:

$$\hat{H} - \hat{\rho}\hat{H}\hat{\rho} \quad (6.13)$$

is used to form the new projected Hamiltonian. However, projecting the Hamiltonian in this manner leads to an energy spectrum where all the valence energies are equal to zero, which is only desirable when all the conduction energies are negative and so more favourable in energy than the zeroed valence states. To avoid this problem the energy spectrum is shifted so that all the valence states become higher in energy than the required conduction states. To guarantee that this shifting is sufficient, the shift must be greater than or equal to the highest conduction energy. Conjugate gradients can easily be used to find a single extremal eigenvalue, so we can use it to find the highest energy eigenvalue within the current conduction NGWF basis. The shift must therefore be greater than this eigenvalue. The projected Hamiltonian can be modified to include the shift, σ, so that the final operator is:

$$\hat{H} - \hat{\rho}\left(\hat{H} - \sigma\right)\hat{\rho}. \tag{6.14}$$

In practice, the shift σ is set to be higher than the highest conduction energy, so that in general it remains constant even when there are changes in the highest eigenvalue, adding stability to the minimization process. If necessary, it can also be updated during the calculation.

6.1.4 Results and Discussion

These five methods were tested and compared for a system with a Kronig-Penney potential using a block update preconditioned conjugate gradients method [11]. By applying the appropriate level of preconditioning and selecting good choices for the reference energies, the results in Table 6.1 were obtained. No shift was applied for the projection method. In attempting to choose good values for the reference energies, it was verified that a poor choice can result in significantly slower convergence. For all of the methods the total conduction energies calculated were accurate to within 10^{-10} Ha of the correct result.

The results show that the different methods are fairly similar in terms of both speed and accuracy, with the projection method as the clear favourite. An important

Table 6.1 Results for the different conduction methods, showing averages for time taken and the number of iterations for a total of 100 calculations with randomly generated starting guesses for the eigenvectors.

Method	Avg. time taken (s)	Avg. number of iterations
Folded spectrum	2.39	182
Shift invert +	2.34	158
Shift invert −	2.23	170
Shift invert i	5.48	463
Projection	1.21	36

Shift invert +, − and i refer to the three variants of the shift invert method discussed in the text. The first form of the projection method was applied, without the use of a shift requirement of the selected method is the need for linear-scaling. Whilst this is hard to test within this basic implementation due to the lack of localization and sparse matrix multiplication, it can be shown that with the appropriate level of preconditioning, the number of iterations required for increasing system size remains approximately constant for the projection method. Combined with the fact that the method mainly consists of matrix multiplications, it seems likely that favourable scaling could be achieved when implemented within local-orbital methods such as ONETEP.

The reason for the relatively large number of iterations required for the folded spectrum method can be seen by considering the condition number, which will be higher for the folded Hamiltonian. Using the approximate expression [12]:

$$\kappa \approx \frac{(\mathcal{E}_{\max} - \mathcal{E}_{\min})}{\mathcal{E}_{\mathrm{gap}}}, \tag{6.15}$$

it is clear that the largest eigenvalue $\mathcal{E}_{\max}$ will be much bigger for the transformed Hamiltonian, and thus so will the condition number, κ. Therefore when using an iterative minimization scheme, convergence will be slower compared to solving the original equation.

For both the shift invert and folded spectrum methods, the choice of reference energy is particularly important. For the folded spectrum method, for example, if it is too low then unwanted valence states will be re-calculated, if it is too high then unwanted high energy conduction states will need to be calculated in order to get the lowest conduction states. Additionally a poor choice of reference energy will result in slower convergence for the shift invert method. For example, if the reference energy is too close to a given eigenvalue, such that the difference between E_{ref} and $\mathcal{E}$ is very small compared to the distance to other eigenvalues, the magnitude of the eigenvalue for the new system will be much greater than all other eigenvalues. This will result in a high condition number, so care must be taken to find a good reference energy. The projection method, however, has the advantage that no reference energy is required and therefore it is more automatic. Additionally, the density matrix is already calculated within a self-consistent ground state local-orbital calculation and so can easily be reused.

For the case of all three methods, the accuracy of the conduction states will clearly be affected by the accuracy with which the potential has been calculated. However, the projection method will also be affected by the accuracy of the valence density matrix, whereas the folded spectrum and shift invert methods will not. This will be particularly significant when the localization and truncation approximations required for linear-scaling behaviour are applied.

6.1.5 Expanding the Projection Method

The methods outlined above were applied directly to the solution of an eigenvalue equation. However in a ONETEP calculation, the system is solved using a density matrix scheme, within the representation of a basis of NGWFs. It is therefore necessary to adapt the methods described above for use within this context. As the projection method has proven to be the most favourable, this is the one which has been further developed.

The general scheme for the projection method within the test program begins with some initial ground state calculation analogous to that of ONETEP, incorporating a set of NGWFs and a density matrix representation. Once the ground state calculation has then been completed, some initial guess is formed for the conduction NGWFs, which can either be random or an arbitrary set of Gaussian functions. From this initial guess, the overlap, inverse overlap and Hamiltonian matrices are then generated by combining the matrix elements from the underlying basis, in this case B-splines, which for the overlap matrix for example would involve the following:

$$S_{\alpha\beta} = \langle \phi_\alpha | \phi_\beta \rangle \tag{6.16}$$

$$= d_\alpha^i \langle b_i | b_j \rangle d_\beta^j \tag{6.17}$$

$$= d_\alpha^i d_\beta^j s_{ij}, \tag{6.18}$$

where Greek indices are used to denote the NGWF basis, Latin indices are used to denote the underlying basis functions $\{|b_i\rangle\}$ and the Einstein summation convention has again been used. The NGWFs are assumed to be real, and are related to the underlying basis via:

$$|\phi_\alpha\rangle = d_\alpha^i |b_i\rangle. \tag{6.19}$$

Before the Hamiltonian is generated, however, it must first be projected with $\hat{I} - \hat{\rho}$, to effectively set all the eigenvalues associated with valence states to zero. Unlike the basic adaptation used to calculate the results described in Sect. 6.1.4, however, the aim is to only calculate the conduction states, rather than calculating the valence and conduction states then projecting out the valence contribution. Therefore, the shift must be applied to the Hamiltonian as described in Eq. 6.14. Within the test program, this was initially achieved by using conjugate gradients to find the highest eigenvalue, λ, of the underlying basis and therefore defining the shift as $\sigma = \lambda + \delta$, where δ is some small additional shift to ensure the highest conduction eigenvalue becomes negative. The underlying Hamiltonian was then shifted such that $h'_{ij} = h_{ij} - \sigma s_{ij}$. The Hamiltonian of the underlying basis can then be projected, so that when the Hamiltonian of the NGWF basis is calculated from the underlying projected Hamiltonian, it will automatically be projected appropriately. In fact, this method of projection corresponds to a shifting down of the conduction states, rather than a shifting up of the valence states, i.e. the following Hamiltonian operator is used:

$$\left(\hat{H} - \sigma\right) - \hat{\rho}\left(\hat{H} - \sigma\right)\hat{\rho}. \tag{6.20}$$

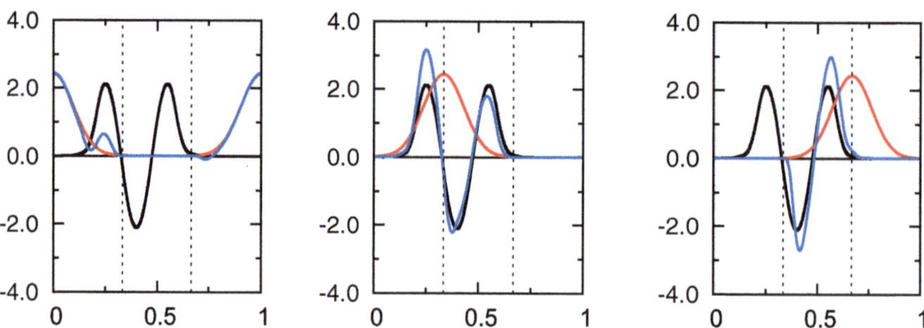

Fig. 6.2 Demonstration of NGWF optimization in the test program, showing the change for a system with 3 NGWFs calculating the eigenstate shown in *black*. The NGWFs start as Gaussian functions [*red*] and adapt in such a manner that the final NGWFs [*blue*] resemble the eigenstate within their localization regions, the boundaries of which are indicated by the *dotted black lines*

This was selected within the test program due to the advantage of requiring the calculation of the largest eigenvalue and the projection of the Hamiltonian only once, however this would be far less practical within the large basis of psincs used in ONETEP. Instead, the Hamiltonian in the conduction basis must be projected directly, as described in the following section. In such a situation the original description of the shift is more advantageous, due to a reduced number of matrix operations required and a more intuitive definition of the total conduction energy.

To further imitate the behaviour of ONETEP, the NGWFs in the test program were localized within some user specified region, and truncation of the density kernel was also included. Figure 6.2 shows an example of NGWF optimization in the test program; the NGWFs were initialized to Gaussian functions within their localization regions and following the energy minimization process they approximately resemble the eigenfunction they are aiming to represent.

As with ONETEP, a combination of the purification transformation, canonical purification and the LNV method were used to impose idempotency and minimize the energy with respect to the DM. Using the above combination of methods, a band structure was calculated for a Kronig-Penney potential. As with ONETEP calculations, the valence NGWF basis functions were optimized in such a manner as to be able to accurately represent those states designated as valence states, however they were not able to accurately reproduce the correct band structure in the designated conduction regime, as shown in Fig. 6.3. However, using the projection method to optimize a set of conduction NGWFs it was possible to achieve excellent agreement with the correct result. Following the success of the implementation in the toy model, the projection method was then implemented in ONETEP, as described in the following section.

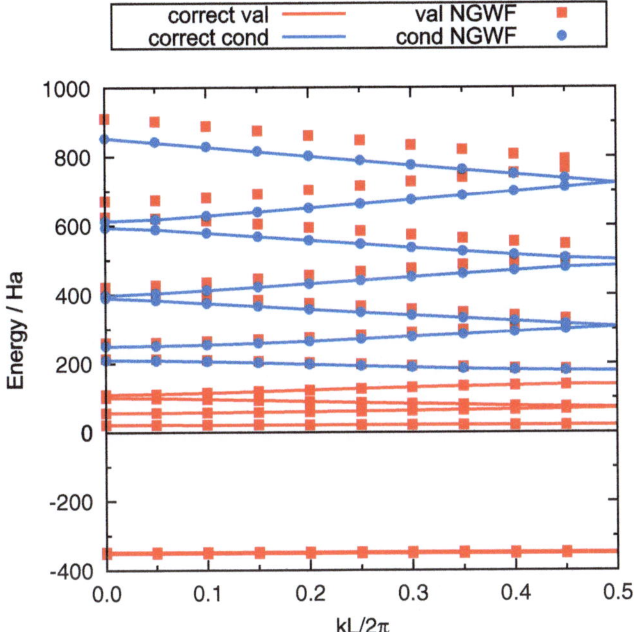

Fig. 6.3 A band structure calculated for the Kronig-Penney potential. The correct result (calculated using direct diagonalization in the B-spline basis) for the valence (conduction) regime is indicated by the *red* (*blue*) *lines*. The results obtained using NGWFs optimized for the valence (conduction) states are plotted as *red squares* (*blue circles*). As with ONETEP, the valence regime accurately reproduces the valence band structure and the lower lying conduction states, but fails to accurately represent the higher energy conduction states. Using the projection method, the conduction NGWFs are able to successfully reproduce the band structure in the conduction regime

6.2 Implementation in ONETEP

6.2.1 Calculating Conduction States

Two sets of NGWFs are now required, $\{|\phi_\alpha\rangle\}$ for the valence states, and $\{|\chi_\alpha\rangle\}$ for the conduction states. The ground state ONETEP calculation already provides access to the valence DM ρ and kernel $\mathbf{K}$, overlap matrix $\mathbf{S}_\phi$ and Hamiltonian $\mathbf{H}_\phi$. The additional conduction matrices will be labelled as follows: $\mathbf{S}_\chi$ is the conduction overlap matrix, $\mathbf{T}$ is the rectangular valence-conduction cross overlap matrix defined as $T_{\alpha\beta} = \langle\phi_\alpha|\chi_\beta\rangle$, $\mathbf{H}_\chi$ is the (unprojected) conduction Hamiltonian, $\mathbf{H}_\chi^{\text{proj}}$ is the projected conduction Hamiltonian, $\mathbf{Q}$ is the conduction DM and $\mathbf{M}$ is the conduction density kernel. These are all represented by atom-blocked sparse matrices [13, 14], such that all matrix-matrix operations are possible in asymptotically linear-scaling computational effort, due to the strict truncation.

The final expression for the projected conduction Hamiltonian, including the shift, σ, is therefore defined as:

$$
\begin{aligned}
\left(H_\chi^{\text{proj}}\right)_{\alpha\beta} &= \langle\chi_\alpha|\hat{H} - \hat{\rho}\left(\hat{H} - \sigma\right)\hat{\rho}|\chi_\beta\rangle \\
&= (H_\chi)_{\alpha\beta} - \left(T^\dagger K H_\phi K T\right)_{\alpha\beta} + \sigma\left(T^\dagger K S_\phi K T\right)_{\alpha\beta}
\end{aligned}
\tag{6.21}
$$

The energy expression $E = \text{Tr}\left[\mathbf{M}\mathbf{H}_\chi^{\text{proj}}\right]$ can then be minimized by optimizing both the set of conduction NGWFs and the conduction kernel. This follows the same general procedure as a ground state ONETEP calculation, without the need for self-consistency or updating of the density and potential. However, extra terms will be needed in the gradients for both the NGWF and kernel optimization.

The total energy functional, including the LNV enforcement of idempotency, can be written:

$$
\Omega = \text{Tr}\left[3\tilde{\mathbf{M}}\mathbf{S}_\chi\tilde{\mathbf{M}}\mathbf{H}_\chi^{\text{proj}} - 2\tilde{\mathbf{M}}\mathbf{S}_\chi\tilde{\mathbf{M}}\mathbf{S}_\chi\tilde{\mathbf{M}}\mathbf{H}_\chi^{\text{proj}}\right]
\tag{6.22}
$$

$$
= \text{Tr}\left[\left(3\tilde{\mathbf{M}}\mathbf{S}_\chi\tilde{\mathbf{M}} - 2\tilde{\mathbf{M}}\mathbf{S}_\chi\tilde{\mathbf{M}}\mathbf{S}_\chi\tilde{\mathbf{M}}\right)\left(\mathbf{H}_\chi - \mathbf{T}^\dagger\mathbf{K}\mathbf{H}_\phi\mathbf{K}\mathbf{T} + \sigma\mathbf{T}^\dagger\mathbf{K}\mathbf{S}_\phi\mathbf{K}\mathbf{T}\right)\right],
\tag{6.23}
$$

where we have defined an auxiliary kernel, $\tilde{\mathbf{M}}$, so that $\mathbf{M} = 3\tilde{\mathbf{M}}\mathbf{S}\tilde{\mathbf{M}} - 2\tilde{\mathbf{M}}\mathbf{S}\tilde{\mathbf{M}}\mathbf{S}\tilde{\mathbf{M}}$. The derivative with respect to the conduction NGWFs is therefore:

$$
\begin{aligned}
\frac{\partial\Omega}{\partial\langle\chi_\gamma|} =\; &3|\chi_\beta\rangle\left(\tilde{M}H_\chi^{\text{proj}}\tilde{M}\right)^{\beta\gamma} - 2|\chi_\beta\rangle\left(\tilde{M}S_\chi\tilde{M}H_\chi^{\text{proj}}\tilde{M}\right)^{\beta\gamma} \\
&- 2|\chi_\beta\rangle\left(\tilde{M}H_\chi^{\text{proj}}\tilde{M}S_\chi\tilde{M}\right)^{\beta\gamma} \\
&+ 3\hat{H}|\chi_\beta\rangle\left(\tilde{M}S_\chi\tilde{M}\right)^{\beta\gamma} - 2\hat{H}|\chi_\beta\rangle\left(\tilde{M}S_\chi\tilde{M}S_\chi\tilde{M}\right)^{\beta\gamma} \\
&- 3|\phi_\beta\rangle\left(K H_\phi K T\tilde{M}S_\chi\tilde{M}\right)^{\beta\gamma} + 2|\phi_\beta\rangle\left(K H_\phi K T\tilde{M}S_\chi\tilde{M}S_\chi\tilde{M}\right)^{\beta\gamma} \\
&+ 3\sigma|\phi_\beta\rangle\left(K S_\phi K T\tilde{M}S_\chi\tilde{M}\right)^{\beta\gamma} - 2\sigma|\phi_\beta\rangle\left(K S_\phi K T\tilde{M}S_\chi\tilde{M}S_\chi\tilde{M}\right)^{\beta\gamma},
\end{aligned}
\tag{6.24}
$$

which can more succinctly be written as:

$$
\frac{\partial\Omega}{\partial\langle\chi_\gamma|} = \hat{H}|\chi_\beta\rangle M^{\beta\gamma} + |\chi_\beta\rangle Q^{\beta\gamma} - |\phi_\beta\rangle\left(K H_\phi K T M\right)^{\beta\gamma} + \sigma|\phi_\beta\rangle\left(K S_\phi K T M\right)^{\beta\gamma},
\tag{6.25}
$$

with:

$$
\mathbf{Q} = 3\tilde{\mathbf{M}}\mathbf{H}_\chi\tilde{\mathbf{M}} - 2\tilde{\mathbf{M}}\mathbf{S}_\chi\tilde{\mathbf{M}}\mathbf{H}_\chi\tilde{\mathbf{M}} - 2\tilde{\mathbf{M}}\mathbf{H}_\chi\tilde{\mathbf{M}}\mathbf{S}_\chi\tilde{\mathbf{M}}.
\tag{6.26}
$$

The tensorially correct gradient can then be found via:

$$g_\gamma = \frac{\partial \Omega}{\partial \langle \chi_\alpha |} \left(S_\chi \right)_{\alpha\gamma}.$$ (6.27)

An additional factor of 2 is required to account for the NGWFs being real and in practice we actually need the gradient with respect to the NGWF coefficients, which can be easily found from the above expression.

The gradient with respect to the (auxiliary) conduction kernel is:

$$\frac{\partial \Omega}{\partial \tilde{M}^{\gamma\beta}} = 6 \left(S_\chi \tilde{M} H_\chi^{\mathrm{proj}} + H_\chi^{\mathrm{proj}} \tilde{M} S \right)_{\beta\gamma}$$
$$- 4 \left(S_\chi \tilde{M} S_\chi \tilde{M} H_\chi^{\mathrm{proj}} + S_\chi \tilde{M} H_\chi^{\mathrm{proj}} \tilde{M} S_\chi + H_\chi^{\mathrm{proj}} \tilde{M} S_\chi \tilde{M} S_\chi \right)_{\beta\gamma},$$ (6.28)

which has the same form as for the valence kernel gradient. This must also be modified to find the tensorially correct contravariant search direction:

$$G^{\gamma\beta} = S_\chi^{\gamma\delta} \frac{\partial \Omega}{\partial \tilde{M}^{\delta\alpha}} S_\chi^{\alpha\beta}.$$ (6.29)

Diagonalization

Once the set of conduction NGWFs has been optimized, one has a choice of Hamiltonians which can be diagonalized to obtain the eigenvalue spectrum. This can be either the unprojected Hamiltonian in the conduction NGWF basis or the projected Hamiltonian. Alternatively, the most useful approach is to diagonalize the Hamiltonian in a joint basis of valence and conduction NGWFs, which gives an improved eigenvalue spectrum. Whilst there is a small additional cost associated with diagonalizing the Hamiltonian in a larger basis of NGWFs, this has the particular advantage that it allows eigenvalues and other properties to be calculated in a basis that is capable of representing both the valence and conduction states of the system. Additionally, this proves to be particularly important when calculating band structures, as discussed in Sect. 7.3.1.

6.2.2 Avoiding Local Minima

It has been observed that it is sometimes possible to become trapped in a local minimum when optimizing the conduction orbitals. This behaviour is characterized by slow convergence of the conduction NGWFs, wherein the RMS gradient stagnates or increases while the energy continues to decrease; or by sharp jumps in the energy with increasing conduction NGWF radii, rather than the expected smooth

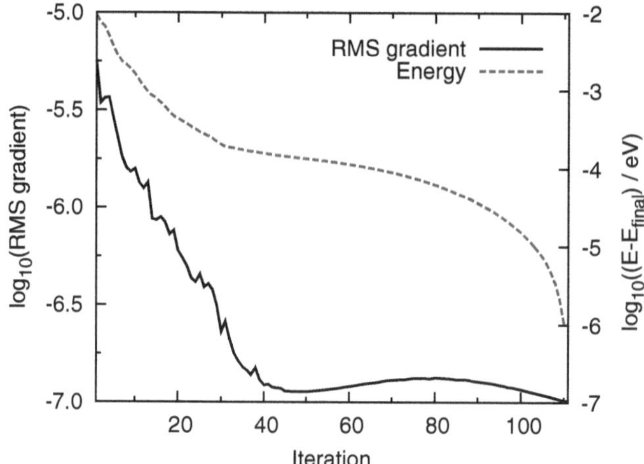

Fig. 6.4 Demonstration of the appearance of a local minimum, where the RMS gradient increases for a period whilst the energy continues to decrease. This was for a geometry-optimized structure of metal-free phthalocyanine at a radius of 18 Bohr with four extra states being optimized

convergence. Examples of both of these features can be seen in Figs. 6.4 and 6.5 respectively. This has been seen to occur due to an unfavourable ordering of the energy eigenstates in the unoptimized basis of NGWFs, so that the NGWFs are optimized for some eigenstates which will eventually be higher in energy at the expense of those which will eventually be lower in energy.

Whilst the occurrence of such behaviour is strongly system dependent, we have developed a method which can be used to overcome the problem. Unlike a ground state calculation where the number of occupied states is fixed by the number of electrons in the system, one is free to choose the number of conduction states one wishes to optimize in a conduction calculation and so this number can easily be varied. Therefore, instead of optimizing for the required number of states from the start of the calculation, we introduce an additional "pre-optimization" stage wherein the conduction NGWFs are initially optimized for a greater number of conduction states than required for a small number of conjugate gradient iterations. The number of states is then reduced to that actually required, the conduction density kernel is regenerated accordingly and the calculation then proceeds to full convergence of the conduction NGWFs using the results of the initial stage as the new starting guess.

The first stage of this process aims to overcome the problem of poor initial ordering of states, whilst the second stage will allow for closer optimization of those states actually required. This is illustrated by Table 6.2, where the LUMO+14 state is initially much higher in energy so that if no additional states are included the NGWFs are not optimized to represent it and it ends up significantly higher in energy than other states. If, however, four additional states are included, this is sufficient to reorder the states and it becomes lower in energy.

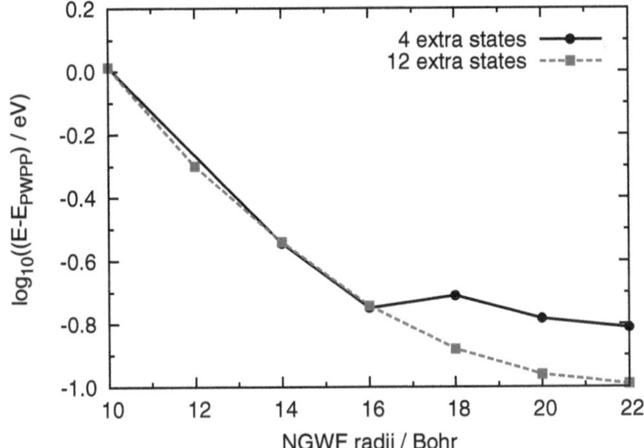

Fig. 6.5 Comparison of conduction energy convergence with respect to conduction NGWF radii for different numbers of extra states, for a geometry-optimized structure of metal-free phthalocyanine. The energy difference is calculated with respect to the traditional PWPP result. A discontinuity appears in the curve at 18 Bohr when four additional states are optimized, demonstrating problems with local minima, whilst optimizing 12 additional states is sufficient to overcome the problem

Table 6.2 Initial energies and values after 5 iterations both with and without optimizing extra states for three different eigenstates of a geometry-optimized structure of metal-free phthalocyanine with conduction NGWF radii of 14 Bohr.

State	Initial	0 extra states	4 extra states
LUMO+14	0.628	>0.368	**−0.042**
LUMO+15	**0.355**	**0.045**	**0.039**
LUMO+16	**0.259**	**0.082**	0.061

States shown in bold are those which are among the 16 lowest states, and thus being included in the conduction NGWF optimization. Without the optimization of extra states, the LUMO+14 state is not optimized and so remains high in energy, whilst with the addition of four extra states, the correct order is found and the LUMO+14 is significantly lowered in energy

The application of this scheme must be done systematically as the appearance of such problems vary from system to system, as does the number of extra states required to resolve the problem. It is therefore very important to perform careful convergence tests to ensure that the calculated conduction states do not correspond to any local minima. This will require variation of the number of conduction NGWFs per atom, convergence with respect to conduction NGWF radii, and an increase in the number of extra conduction states requested, until consistent results are achieved, with a smooth curve of energy against NGWF radii, and sensible convergence of the NGWFs during a calculation. By following these strategies one can become confident that accurate results have been achieved.

6.2.3 Band Structures

Both methods of band structure calculation had been previously implemented in ONETEP, with the necessary extensions to three dimensional systems and the use of non-local potentials. As the calculation takes place as a post-processing step, once either or both sets of NGWFs are fully optimized, it is possible to use the joint basis of valence and conduction NGWFs directly. As will be seen in Sect. 7.3.1, the use of the joint basis is vital for the accurate calculation of the band structure across the Brillouin zone.

One restriction to the use of the tight-binding style method is that the basis functions used must, of course, be well localized. In ONETEP, particularly for conduction calculations, it is possible, although not desirable to use large NGWF radii such that the NGWFs extend over a large portion of the unit cell. Whilst this is not a problem for the largest supercells used, for smaller systems there will be some restriction on the applicability of the tight-binding style method when large NGWF radii are used. More precisely, the NGWF radii must be small enough that no NGWFs centred in the unit cell overlap with any NGWFs associated with its periodic image. This is equivalent to the condition that must be fulfilled to allow the application of the position operator to molecular systems.

6.2.4 Application to Spectroscopy

We follow the method for the calculation of optical absorption spectra as applied in the CASTEP PWPP code [15], as described in Sect. 4.2.2. As with the calculation of band structures, this calculation occurs after the NGWF optimization process, and so again it is possible to use the joint valence and conduction NGWF basis.

In principle the expression for the imaginary component of the dielectric function (Eq. 4.26) includes a $\mathbf{k}$-point sum over the entire Brillouin zone, however as with ground state ONETEP calculations, it is assumed that a large enough supercell will be used such that only the Γ-point need be considered. This could be extended in future using either of the two methods for interpolating band structures in ONETEP described in the previous Chapter. For the purposes of this work, however, all calculations have been restricted to the Γ-point only.

As discussed in Sect. 4.2.1, the position operator is not well defined in periodic boundary conditions, and so one must instead use a momentum representation to calculate the transition matrix elements required for the calculation of optical absorption spectra. This necessitates the calculation of the commutator between the non-local potential and the position operator. In ONETEP this is calculated using the following identity [16]:

$$(\nabla_{\mathbf{k}} + \nabla_{\mathbf{k}'}) \left[\int e^{-i\mathbf{k}\cdot\mathbf{r}} V_{\mathrm{nl}}(\mathbf{r}, \mathbf{r}') e^{i\mathbf{k}'\cdot\mathbf{r}'} d\mathbf{r}\, d\mathbf{r}' \right]$$

$$= i \int e^{-i\mathbf{k}\cdot\mathbf{r}} \left[V_{\mathrm{nl}}(\mathbf{r}, \mathbf{r}')\, \mathbf{r}' - \mathbf{r} V_{\mathrm{nl}}(\mathbf{r}, \mathbf{r}') \right] e^{i\mathbf{k}'\cdot\mathbf{r}'} d\mathbf{r}\, d\mathbf{r}', \tag{6.30}$$

where the derivative can be calculated using finite differences in reciprocal space. Alternatively, it can be determined analytically, which requires the derivative of the angular momentum projectors of the non-local pseudopotential. These are defined in reciprocal space as:

$$\chi_{lm}^{(I)}(\mathbf{k}) = e^{i\mathbf{k}\cdot\mathbf{R}_{(I)}} 4\pi (-i)^l Z_{lm}(\Omega_{\mathbf{k}}) \zeta_l^{(I)}(k), \tag{6.31}$$

where the Z_{lm} are real spherical harmonics and $\mathbf{R}_{(I)}$ is the position vector of atom I. The derivative of the spherical harmonic can be determined analytically, however the derivative of the radial part of the projectors, $\zeta_l^{(I)}(k)$, must be interpolated to allow the calculation of its derivative using a finite difference approach.

The matrix elements are calculated in this manner and used to form a weighted density of states, which is smeared using Gaussian functions. We note that it is possible to calculate position matrix elements between eigenstates for molecules within periodic boundary conditions, providing the NGWF radii are sufficiently small that no NGWFs associated with the molecule overlap with any NGWFs associated with its periodic image. This is equivalent to the condition for the use of the tight-binding style method for band structure calculation, as described in the previous section.

Once the transition matrix elements and eigenvalues have been successfully calculated in ONETEP, they can then be used to calculate the imaginary component of the dielectric function. This is done in a manner equivalent to the calculation of DOS, where the eigenvalues are smeared using Gaussian functions. At this stage, the scissor operator approximation can also be applied to facilitate comparison with experiment.

6.3 Summary

In this chapter we have presented three different methods for the accurate calculation of the unoccupied Kohn-Sham states. These were applied within a toy model and the projection method was consequently selected as the most efficient. This method was further developed for application within a local orbital code and the implementation within ONETEP was described, including the details of the application to the calculation of optical absorption spectra. In the following chapter, we present the results of the application of the projection method to molecular and extended systems, demonstrating the effectiveness of the method in calculating DOS, band structures and the imaginary component of the dielectric function.

References

1. Ratcliff, L.E., Hine, N.D.M., Haynes, P.D.: Phys. Rev. B **84**, 165131 (2011)
2. MacDonald, J.K.L.: Phys. Rev. **46**(9), 828 (1934)
3. Wang, L.-W., Zunger, A.: J. Chem. Phys. **100**(3), 2394 (1994)
4. Ericsson, T., Ruhe, A.: Math. Comput. **35**(152), 1251 (1980)
5. Tackett, A.R., Di Ventra, M.: Phys. Rev. B **66**(24), 245104 (2002)
6. Martins, A.S., Boykin, T.B., Klimeck, G., Koiller, B.: Phys. Rev. B **72**(19), 193204 (2005)
7. Fu, H., Zunger, A.: Phys. Rev. B **56**(3), 1496 (1997)
8. Franceschetti, A., Zunger, A.: Phys. Rev. Lett. **78**(5), 915 (1997)
9. Cha, P.D., Gu, W.: J. Sound Vib. **227**(5), 1122 (1999)
10. Stewart, G.W.: Matrix Algorithms Volume 2: Eigensystems. Siam, Philadelphia, 2001
11. Mostofi, A.A., Haynes, P.D., Skylaris, C.-K., Payne, M.C.: J. Chem. Phys. **119**(17), 8842 (2003)
12. Goedecker, S.: Rev. Mod. Phys. **71**, 1085 (1999)
13. Hine, N.D.M., Haynes, P.D., Mostofi, A.A., Skylaris, C.-K., et al.: Comput. Phys. Commun. **180**(7), 1041 (2009)
14. Hine, N.D.M., Haynes, P.D., Mostofi, A.A., Payne, M.C.: J. Chem. Phys. **133**(11), 114111 (2010)
15. Pickard, C.J.: Ab initio electron energy loss spectroscopy. Ph.D. thesis, University of Cambridge (1997)
16. Motta, C., Giantomassi, M., Cazzaniga, M., Gal-Nagy, K., et al.: Comput. Mater. Sci. **50**(2), 698 (2010)

Chapter 7
Results and Discussion

In this chapter we present results calculated in ONETEP using the projection method described in the previous chapter. We apply the method to both molecular and extended systems. For the smaller system sizes these results have been compared to those obtained from a PWPP code, validating the results obtained. We also include results calculated using the valence NGWF basis only, which emphasizes the necessity of the projection method for obtaining accurate DOS, band structures and absorption spectra. For all systems, the local minima avoiding scheme described in Sect. 6.2.2 was used to ensure the conduction states obtained corresponded to the true minimum. We also test the scaling of the method for a conjugated polymer, which proves to be linear for systems containing up to 1000 atoms, which was the maximum system size tested. Finally, we discuss some of the limitations of the method, highlighting the importance of systematic convergence for conduction calculations. All orbital, NGWF and atomic structure plots here and elsewhere were plotted with XCrySDen [1].

Parts of this chapter have been previously published in the following paper: Ratcliff et al., Calculating optical absorption spectra for large systems using linear-scaling density-functional theory, Copyright (2011) by the American Physical Society [2].

7.1 Metal-Free Phthalocyanine

As stated in Sect. 3.4 ONETEP is particularly efficient at treating molecules, and so metal-free phthalocyanine was chosen as a good test system on which to apply the conduction state method. As it contains only 58 atoms, calculations could also be performed in CASTEP, with a corresponding plane-wave/psinc kinetic energy cut-off of 1046 eV and identical norm-conserving pseudopotentials. In all calculations the LDA exchange-correlation functional was used.

L. Ratcliff, *Optical Absorption Spectra Calculated Using Linear-Scaling Density-Functional Theory*, Springer Theses, DOI: 10.1007/978-3-319-00339-9_7,
© Springer International Publishing Switzerland 2013

Phthalocyanines and their derivatives are commonly used as dyes and are also of interest in a number of other fields, including use in photovoltaic cells [3] and molecular spintronics [4] and so metal-free phthalocyanine also provides an interesting test case for the calculation of optical absorption spectra. Therefore in this section we will present DOS and optical absorption spectra results for four different structures of metal-free phthalocyanine.

The atomic coordinates for metal-free phthalocyanine were taken from neutron diffraction data [5] with C_{2h} symmetry and the inner H atoms attached to opposite N atoms. Additional symmetry constraints were then applied by averaging the atomic positions to give the higher symmetry D_{2h} with the inner H atoms attached to both opposite and adjacent N atoms (*trans* and *cis* forms respectively), and finally a geometry optimized structure was calculated using traditional DFT, which was found to have a *trans*-D_{2h} symmetry but differs in bond lengths from the other D_{2h} structure. Diagrams of the *trans* and *cis* forms are shown in Fig. 7.1. Table 7.1 shows the ground state energies for each structure relative to the geometry optimized result, with the higher symmetry structures lower in energy. Very good agreement is achieved between the ONETEP and traditional DFT results.

Conduction States

Following the ground state calculations, a set of conduction NGWFs were optimized for each structure of metal-free phthalocyanine using the projection method. The valence NGWFs were at a fixed radius of 12 Bohr, with one NGWF per H atom, and four each per C and N atom. Sixteen conduction states were optimized, with four conduction NGWFs for each atomic species, and a radius of 16 Bohr was used for the DOS calculations, whilst 13 Bohr was sufficient to achieve almost perfect agreement with traditional DFT for the optical absorption spectrum. This difference in NGWF radii required for good convergence of DOS and optical absorption spectra is discussed in Sect. 7.4. The DOS of the geometry optimized structure is shown in Fig. 7.2, which compares ONETEP results both with and without conduction NGWFs to those found using the PWPP method. For this and all other results in this chapter, we have included the same number of conduction states for both the ONETEP and PWPP results. Without the conduction NGWFs, the ONETEP results differ greatly from the PWPP results, but with the addition of conduction NGWFs, excellent agreement with the PWPP method is achieved.

7.1.1 Optical Absorption Spectra

Optical absorption spectra were then calculated using both the position operator and the momentum operator (including the non-local commutator) for all four structures, and in all cases the two methods agreed almost perfectly with the PWPP results for the energy range considered. The addition of a greater number of conduction states

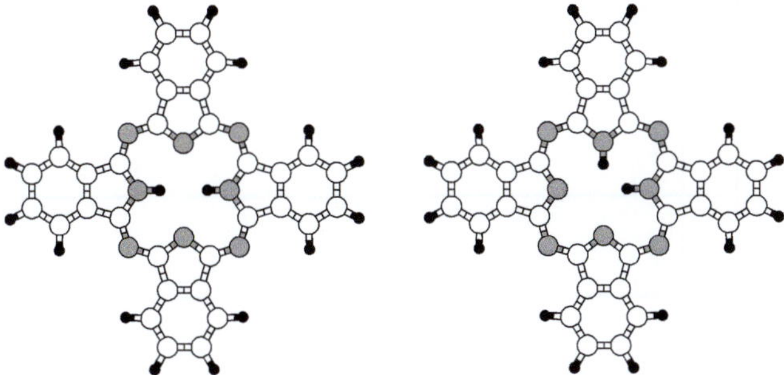

Fig. 7.1 Schematics showing the structures of the *trans* (*left*) and *cis* (*right*) isomers of metal-free phthalocyanine. C atoms are shown in *white*, N atoms in *grey* and H atoms in *black*

Table 7.1 Comparison between ONETEP and PWPP ground state total energies for four different structures of metal-free phthalocyanine, relative to the lowest energy geometry optimized structure

Structure	$E - E_{\text{geom}}$/eV	PWPP
	ONETEP	
C_{2h}	1.554	1.553
cis-D_{2h}	1.875	1.874
trans-D_{2h}	0.952	0.951

The energy difference between the ONETEP and PWPP results for the geometry optimized structure is 0.163 eV, which is due to the slight inequivalence in the basis sets used in the two methods

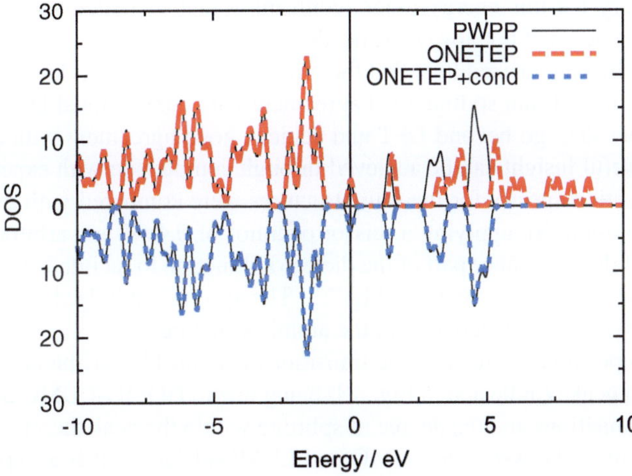

Fig. 7.2 Density of states calculated in both CASTEP and ONETEP for the geometry-optimized structure of metal-free phthalocyanine, plotted with a Gaussian smearing width of 0.1 eV, with a conduction NGWF radius of 16 Bohr. Only the 16 optimized conduction states are included, and the ONETEP+cond curve shown is calculated in the joint valence-conduction NGWF basis

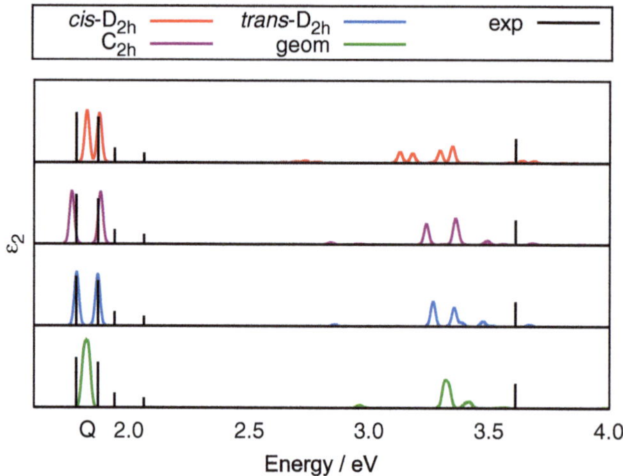

Fig. 7.3 The imaginary component of the dielectric function calculated in ONETEP and plotted for four different structures of metal-free phthalocyanine as indicated on the graph ("geom" refers to the geometry optimized structure). Results from the PWPP method are indistinguishable and so not plotted. A Gaussian smearing width of 0.01 eV is used, with conduction NGWF radii of 13 Bohr and a scissor operator of 0.4 eV. Experimental results in solution (labelled "exp") are also included with the peaks shown as vertical lines for clarity and the calculated results vertically scaled arbitrarily for easier comparison. The position of the Q-bands is indicated below the x-axis

is unnecessary for this energy range, confirming that the calculation of unbound conduction states will not always be needed.

It should be emphasized here that the aim of this work is to calculate absorption spectra within DFT and so find good agreement with conventional DFT implementations, rather than go beyond DFT and achieve good agreement with experiment. However, useful insight can be achieved through comparison with experiment, and so the absorption spectra for the four structures were compared with experimental results in solution [3], applying a scissor operator of 0.4 eV, and arbitrarily scaling the height of the imaginary part of the dielectric function to facilitate easier comparison with experiment, as shown in Fig. 7.3. The spectra are indeed distinguishable, despite the very small differences in the atomic structures.

It is also possible to identify the transitions responsible for the peaks, with the split Q-band peaks (indicated in Fig. 7.3) being due to HOMO-LUMO and HOMO-LUMO+1 transitions and the degree of splitting within the peak therefore due to the energy difference between the LUMO and LUMO+1 bands. It is accepted that the lower symmetry of the metal-free phthalocyanine structure as compared to metal phthalocyanines is the cause for this Q-band splitting, which is not observed for metal phthalocyanines. This agrees with the observation that the higher symmetry *trans*-D_{2h} structure exhibits a lower degree of splitting than the *trans*-C_{2h} structure. The Q-band splitting for the geometry optimized structure is 0.02 eV, which is significantly less than the experimental value of 0.09 eV, implying that the LDA is not sufficiently accurate to calculate the correct structure.

There have already been a number of studies [6–9] of the electronic structure and absorption spectrum of metal-free phthalocyanine, with which the above results are consistent, confirming that this is a useful system to demonstrate the ability of theoretical optical absorption spectra as implemented here to distinguish between similar geometries.

7.2 C_{60}

Calculations were performed for C_{60} at a kinetic energy cut-off of 1115 eV with four valence NGWFs per atom and valence NGWF radii of 10 Bohr. The structure was generated using geometry optimization in CASTEP.

Conduction States

Figure 7.4 shows the density of states for C_{60}, calculated with five conduction NGWFs per atom at a radius of 14 Bohr. As with the other systems studied, excellent agreement is achieved with the PWPP method for those conduction states which have been optimized. Figure 7.5 shows the convergence of the DOS in the conduction regime with respect to the conduction NGWF radii. Those states lower in energy are well converged even at a radius of 10 Bohr, however the higher energy states require a much larger radius in order to achieve a similar level of agreement with the PWPP method.

7.3 Conjugated Polymers

As well as tests on molecular systems, we have applied the method to two different conjugated polymers (shown in Fig. 7.6) and a doped system. These are periodic systems, but retain the advantage of having low dimensionality and so there will be a favourable crossover point at which ONETEP calculations become cheaper than corresponding PWPP calculations. For these systems, the existence of a momentum operator formulation is essential, as discussed in Sect. 4.2.2.

Conjugated Polymers: Discovery, Properties and Applications

Since the discovery in 1976 that conjugated polymers could be made conducting, they have been the subject of much research, both for reasons of fundamental theoretical interest and due to the promise of a wide range of exciting applications. They

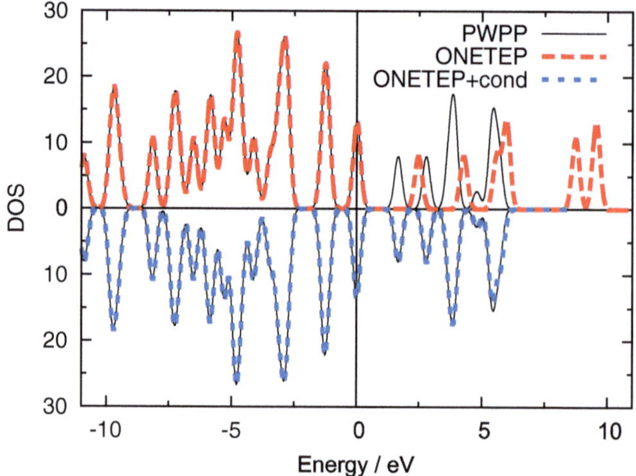

Fig. 7.4 DOS of C_{60} calculated with both CASTEP and ONETEP with conduction NGWF radii of 14 Bohr and a Gaussian smearing width of 0.15 eV. Only the 23 optimized conduction states are plotted, and the ONETEP+cond curve is from the joint valence-conduction NGWF basis

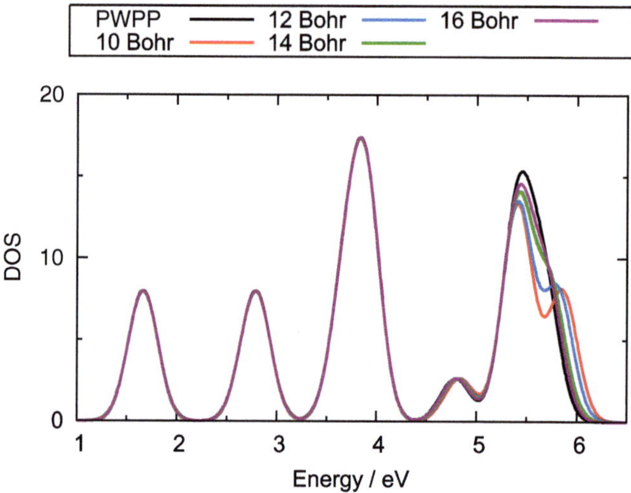

Fig. 7.5 Convergence of the DOS of C_{60} with respect to conduction NGWF radii, compared to the PWPP result

benefit from a variety of unique properties which make them a highly attractive candidate for use in applications ranging from light-emitting diodes (LEDs) [10–12] to photovoltaic cells [12]. Conjugated polymers are particularly suited for applications and theoretical research such as this, in part due to the ability to influence their properties via doping and modification of side chains, enabling for example the creation of materials with a wide range of energy gaps resulting in materials ranging from metallic to insulating [13, 14]. Furthermore, they benefit from useful mechanical properties, such as the ability to be flexible and from desirable processing methods, particularly the ability to process from solution [12, 14, 15]. As a result, a great deal of effort has been extended into developing and studying such polymers with the hopes of replacing inorganic semiconductors for a number of applications.

One of the chief uses for conjugated polymers has been in LEDs, with two of the main polymers used being poly(*para*-phenylene) (PPP) and poly(*para*-phenylene vinylene) (PPV) and their derivatives [10]. In particular blue electroluminescence has been demonstrated using PPP [10, 16]. The ability to tune the band gap has enabled the LED colour to be varied across the visual spectrum with high performance LEDs having been demonstrated [12].

Moreover, conjugated polymers are useful for a number of other optoelectronic devices [15], which in addition to those mentioned previously include photodetectors, optocouplers and light-emitting electrochemical cells (LECs) [12], as well as molecular information storage [14], lasers, batteries, thin-film transistors and all-polymer integrated circuits [15]. Extensive progress has been made towards the development of electroluminescence displays [12] and two-colour LECs have even been fabricated, where the colour is changed when the bias voltage is reversed [17]. This abundance of novel applications has led to an unsurprising growth in the field of plastics electronics. Indeed, the importance of conjugated polymers has been recognized through the granting of the 2000 Nobel prize for chemistry to Heeger, McDiarmid and Shirakawa for their discovery and development.

Conjugated Polymers and Solar Cells

For the purposes of this work, we are particularly interested in the application to photovoltaic cells, where conjugated polymers such as PPP and PPV have also been suggested as possible candidates for use as a donor when in combination with an acceptor such as the fullerene C_{60} [12, 18]. For this use, certain criteria must be met including the occurrence of charge transfer, which has been demonstrated to occur in fullerene-polymer mixtures [19]. An example of this was in C_{60}-doped PPP: Lane et al. [16] compared photoinduced absorption spectra of PPP with C_{60}-doped PPP, for which their results showed an additional sharp peak in the doped sample that was not present in the spectrum of pure PPP, thus demonstrating the occurrence of charge transfer.

One could envisage constructing polymer/C_{60} devices in a variety of ways with either covalent or non-covalent bonds connecting the polymer to the C_{60}, or by creating thin film devices. Indeed there are a wide variety of structures which have already been realized for C_{60}-doped polymers [20]. We have chosen a simple structure for C_{60}-doped PPP, in order to demonstrate the usefulness of calculating optical absorption spectra in ONETEP of systems of such sizes. In practice, it is likely that chemically substituted PPP or other related conjugated polymers might be used instead of pure PPP, due to the increased solubility that can be achieved in organic solvents, which Lane et al. note is useful for easy fabrication of electronic devices. Similarly, it would not necessarily be possible to easily produce experimentally the structure investigated in this work. Nonetheless, this simplification is justified as a first approximation in that it allows us to meet the intended aim of demonstrating the relevance of this work and potential for application to real world problems.

7.3.1 Poly(**para**-*phenylene*)

Calculations were performed for the conjugated polymer poly(*para*-phenylene). The structure for two unit cells was obtained by performing a geometry optimization in CASTEP using the structure of Ambrosch-Draxl et al. [10] as a starting guess, with the final structure shown in Fig. 7.6(a). A cut-off energy of 1115 eV was found to be necessary for good convergence of the results. All calculations were performed at the Γ point only, with no **k**-point sampling, to allow for direct comparison between the two codes. Ground state calculations were first performed with one NGWF per Hydrogen atom and four NGWFs per Carbon atom and a fixed radius of 10 Bohr.

Conduction States

Conduction calculations were performed using four NGWFs for all atomic species with a fixed radius of 14 Bohr. Calculations were performed for a range of different system sizes, with the number of conduction states calculated being set to include all negative eigenvalues for the smallest system studied (corresponding to two unit cells of PPP) and increased linearly with system size. The density of states was plotted for varying chain lengths of PPP, with the graph for 120 atoms shown in Fig. 7.7. As with metal-free phthalocyanine, excellent agreement with CASTEP is achieved for the conduction calculation.

Scaling

This system was also used to test the scaling of the projection method for calculating conduction states by increasing the chain length and comparing the time taken for the calculations. Figure 7.8 shows the scaling behaviour of ONETEP for the conduction calculation. Neither the valence nor conduction density kernels were truncated, however, the behaviour of ONETEP is shown to be approximately linear up to 1000 atoms, and it is expected that this trend will continue up to larger system sizes.

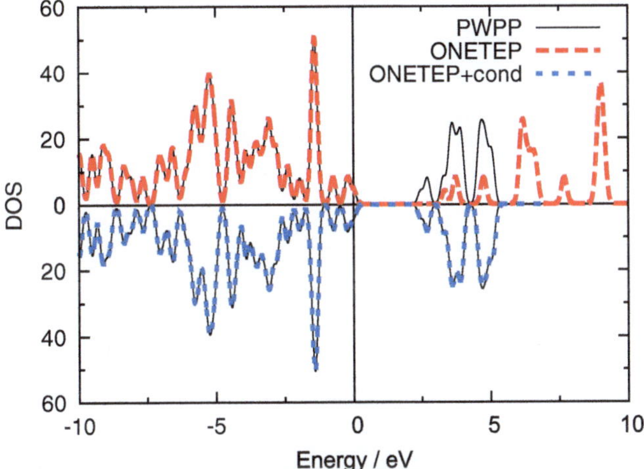

Fig. 7.6 Schematics showing the structures of PPP and PPV, with C atoms shown in *white* and H atoms in *black*. **a** The structure of poly(*para*-phenylene). **b** The structure of poly(*para*-phenylene vinylene)

Fig. 7.7 DOS of 120 atoms of PPP calculated with both CASTEP and ONETEP with conduction NGWF radii of 14 Bohr and a Gaussian smearing width of 0.1 eV. Only the optimized conduction states are included, and the ONETEP+cond curve is from the joint valence-conduction NGWF basis

Optical Absorption Spectrum

The imaginary component of the dielectric function was calculated for varying chain lengths of PPP, using the momentum operator formulation. The result for 120 atoms is shown in Fig. 7.9. As with metal-free phthalocyanine, nearly perfect agreement with CASTEP was achieved with the conduction NGWF basis, whereas the calculation using the valence NGWF basis only showed big discrepancies not only in the positions of the peaks, but also in the relative strengths.

We can also use the system of PPP to demonstrate the possibilities for identifying particular peaks in a given spectrum. Figure 7.10 shows the imaginary part of the dielectric function of PPP split into its three Cartesian components, with the three directions indicated. In addition, the two biggest peaks in the spectrum are highlighted

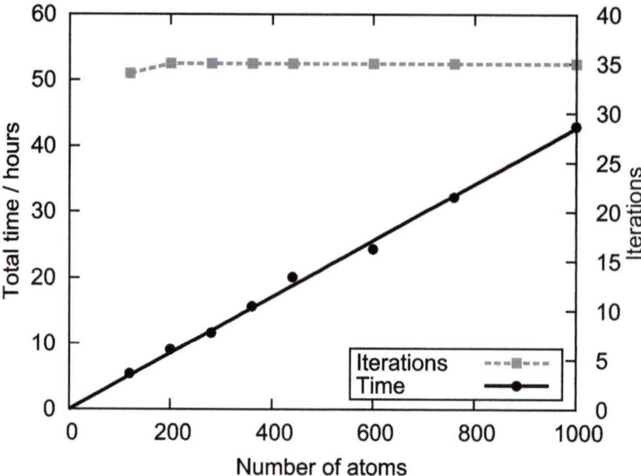

Fig. 7.8 Graph showing the scaling of ONETEP conduction calculations for increasing chain lengths of PPP. Calculations were performed on 8 nodes and therefore a total of 32 cores. The total time taken for ONETEP is approximately linear up to 1000 atoms and the number of NGWF iterations required for convergence (shown on the *right hand* scale) is shown to be constant with an increasing number of atoms

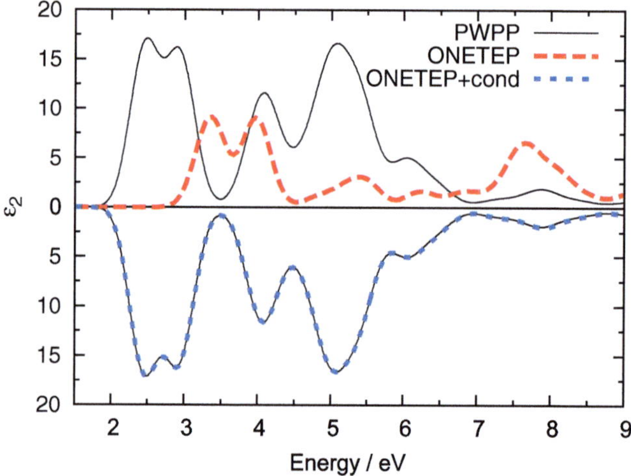

Fig. 7.9 The imaginary component of the dielectric function calculated in CASTEP and ONETEP both with and without a conduction calculation for 120 atoms of PPP. Conduction NGWF radii of 14 Bohr and a Gaussian smearing width of 0.2 eV are used

and the wavefunctions corresponding to the states involved in those transitions are also shown. Both of these transitions are for the component of the dielectric function along the direction of the chain and we can see that the second peak is degenerate.

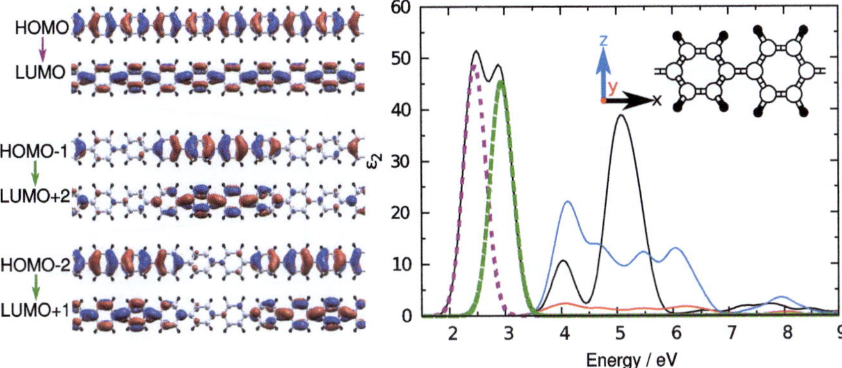

Fig. 7.10 Plot showing the three Cartesian components of the imaginary dielectric function for PPP. The directions are indicated on a schematic of the polymer. In addition, the two lowest energy peaks are highlighted in *green* and *purple* respectively, with the wavefunctions plotted for the states involved in the corresponding transitions on the *left*

Band Structures

The two methods of band structure calculation were also tested for PPP. The results are shown in Fig. 7.11, where band structures have been plotted in the valence NGWF basis, conduction NGWF basis (using the projected Hamiltonian) and the joint valence-conduction NGWF basis. The occupied states are well represented by the valence NGWF basis, however the unoccupied states are not, as would be expected. Only two bands have been plotted in the unoccupied region in the valence NGWF basis and as with the toy model results, we can see that the band has been shifted upwards in energy, with the $\mathbf{k} \cdot \mathbf{p}$ style method preserving the true shape of the band and the tight-binding style method achieving the correct gap between the bands. The results for the conduction NGWF basis show significant improvement over the valence results in the unoccupied region, however whilst the agreement is almost exact at the Γ point, correct results are not achieved throughout the Brillouin zone, particularly using the tight-binding style method. It is only when one combines both sets of NGWFs to form the joint basis that excellent results are achieved throughout the Brillouin zone. When this joint basis is used, the tight-binding style results are marginally better than the $\mathbf{k} \cdot \mathbf{p}$ results, due to its ability to better enforce the correct Brillouin zone periodicity.

7.3.2 Poly(para-phenylene vinylene)

We have also performed calculations on the conjugated polymer poly(*para*-phenylene vinylene), for which the structure was obtained via geometry optimization using the

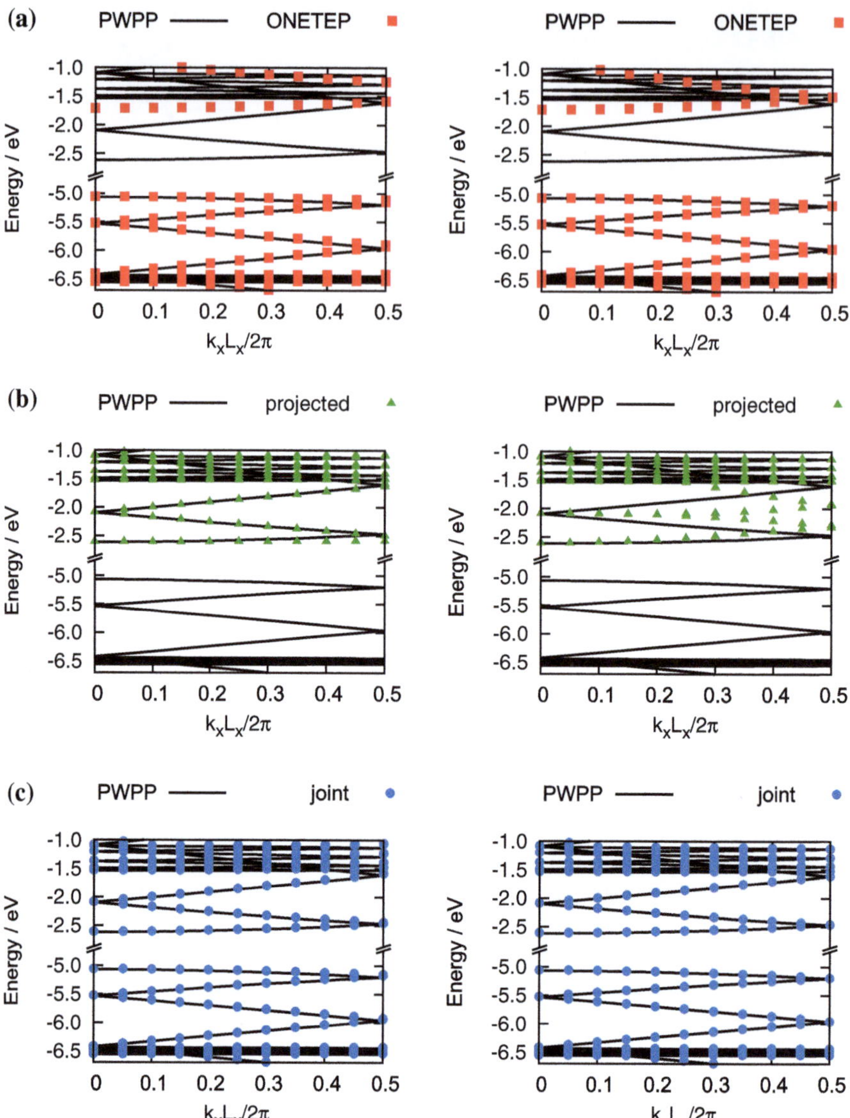

Fig. 7.11 PPP band structures calculated in ONETEP using the $\mathbf{k} \cdot \mathbf{p}$ (*left*) and tight-binding (*right*) style methods in the different NGWF bases are compared to PWPP results. **a** Band structures calculated in the valence NGWF basis. **b** Band structures calculated in the conduction NGWF basis using the projected Hamiltonian. **c** Band structures calculated in the joint valence-conduction NGWF basis

PWPP method starting from the initial structure of Gomes da Costa et al. [21]. A schematic of the final structure is shown in Fig. 7.6b. The PWPP calculations were performed at a kinetic energy cut-off of 1115 eV whilst a cut-off of 1140 eV was used for the ONETEP calculations; any discrepancies between the results due to the different energy cut-offs used are expected to be negligible. A unit cell containing 56 atoms was used and ground state calculations were first performed with one NGWF per Hydrogen atom and four NGWFs per Carbon atom and a fixed radius of 10 Bohr, as for PPP.

Conduction States

Conduction calculations were performed using five NGWFs for all atomic species with a fixed radius of 14 Bohr. The density of states is shown in Fig. 7.12 and excellent agreement between the two methods is again obtained. Figure 7.13 shows the NGWF optimization process for the conduction NGWF basis, with the initial and final NGWFs plotted for a selected H atom.

Optical Absorption Spectrum

The imaginary component of the dielectric function is plotted for PPV in Fig. 7.14, again calculated using the momentum operator, including the non-local commutator

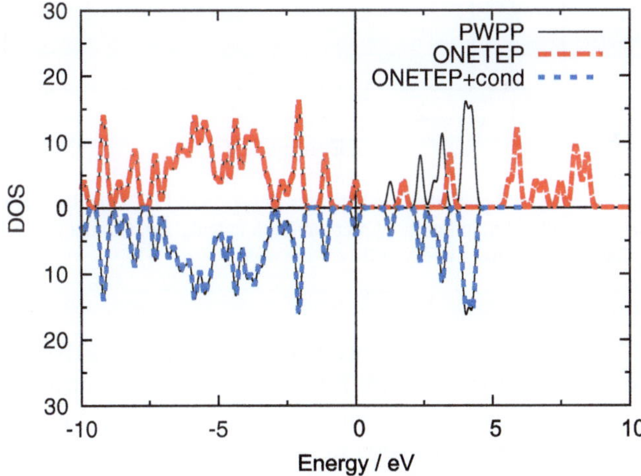

Fig. 7.12 Density of states calculated with both CASTEP and ONETEP for PPV with conduction NGWF radii of 14 Bohr and a Gaussian smearing width of 0.1 eV. Only the fifteen optimized conduction states are plotted, and the ONETEP+cond curve is from the joint valence-conduction NGWF basis

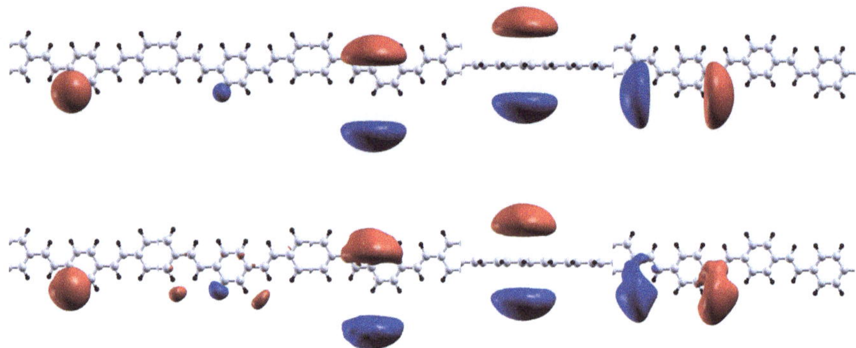

Fig. 7.13 Demonstration of the conduction NGWF optimization process for PPV. The *top row* shows the initial unoptimized conduction NGWFs on a selected H atom and the *bottom row* shows the final optimized conduction NGWFs for the same atom

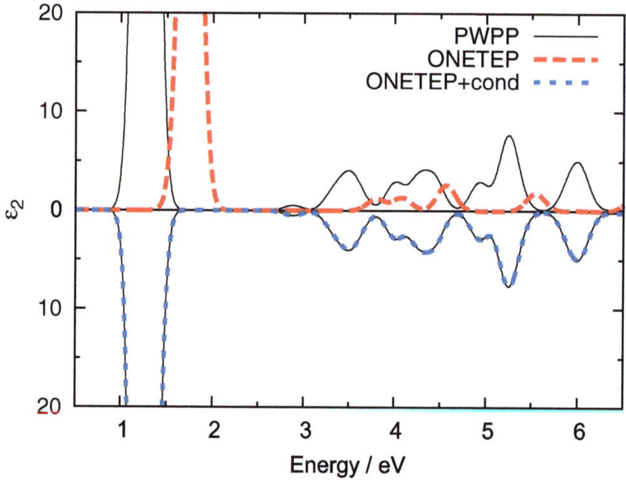

Fig. 7.14 The imaginary component of the dielectric function of PPV calculated using the PWPP method and ONETEP both with and without a conduction calculation. Conduction NGWF radii of 14 Bohr and a Gaussian smearing width of 0.1 eV are used

term. Figure 7.15 shows the convergence with respect to the number of conduction states included, calculated using the PWPP method. For energies up to approximately 6.5 eV the spectrum is remarkably well converged with only 15 conduction states included. This justifies the calculation of bound states only in ONETEP for spectra plotted within the selected energy range. To converge the higher energy part of the spectrum a significant number of conduction states are required, and so caution should be applied to results calculated in ONETEP for this energy region, where states are less likely to be well represented within a localized basis.

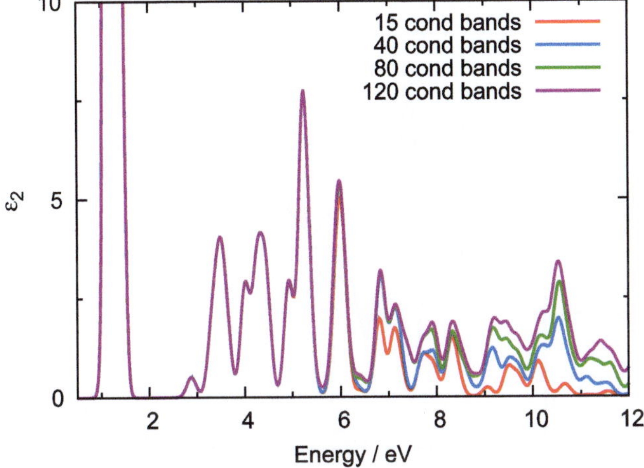

Fig. 7.15 Convergence of the imaginary component of the dielectric function of PPV with respect to the number of conduction states included, calculated using the PWPP method

7.3.3 C_{60}-Doped Poly(para-phenylene)

We have calculated spectra for both PPP and C_{60} individually, as well as a C_{60}-doped poly(*para*-phenylene) chain with a concentration of approximately 25 % by weight, in agreement with that used by Lane et al. This allows us to highlight the benefits of being able to calculate optical absorption spectra in a linear-scaling DFT code, in particular the capability of identifying the origin of various peaks in the spectrum. The bonding structure between the C_{60} molecule and the PPP chain is shown in Fig. 7.16.

Fig. 7.16 Schematic showing the structure of a segment of C_{60}-doped PPP

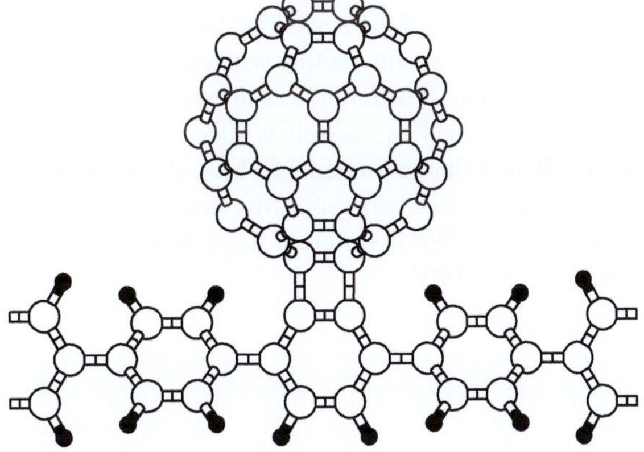

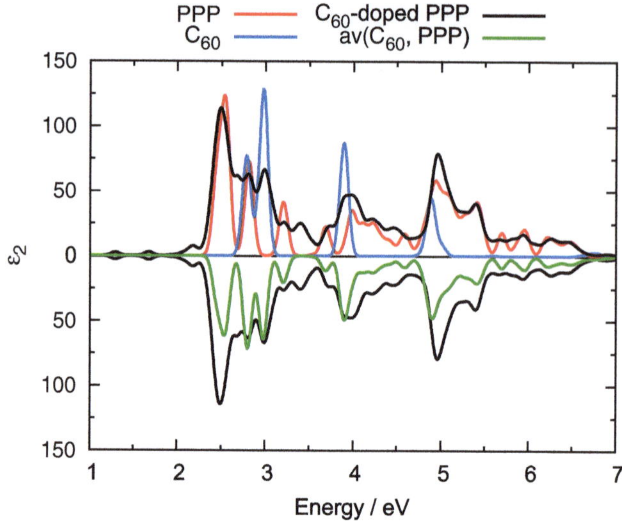

Fig. 7.17 The imaginary component of the dielectric function for C_{60}-doped PPP, compared with that of pure PPP and C_{60}. An average result is also shown for pure PPP and C_{60} for comparison. The results are plotted with a Gaussian smearing width of $0.05\,\text{eV}$

Optical Spectra

Figure 7.17 shows the spectrum for the doped polymer, which is compared with pure PPP, C_{60} and an average of the two individual spectra. A unit cell of 676 atoms was used for the doped polymer, which corresponds to 28 units of PPP and 2 molecules of C_{60}. The spectrum shown for pure PPP is calculated with a unit cell containing 14 units of PPP and for C_{60} the result is for a single molecule, the results are therefore scaled accordingly. The number of NGWFs and their radii were set to the same values as used for C_{60}. The results show a number of small peaks which appear at lower energies than any of the peaks in either of the individual spectra, demonstrating that the PPP and C_{60} are interacting with each other, as one might expect. These peaks are due to a number of different transitions, involving states which are both localized on the chain and the C_{60}, as well as states which are mixed. This agrees with expected results in that charge transfer has been shown to occur in similar systems. One could further explore this behaviour by calculating spectra for systems wherein the C_{60} is attached to the PPP in different ways. In addition, similar results could be generated for C_{60}-doped PPV.

7.4 Discussion and Limitations

As can be seen from the above results, the projection method has proven to be a good method of optimizing a set of NGWFs that are capable of representing the conduction states to a good degree of accuracy. Excellent agreement has been achieved with the PWPP code CASTEP, for the calculation of DOS, band structures and optical absorption spectra. Furthermore, the method has been demonstrated to scale linearly up to system sizes of 1000 atoms. However, there are some limitations to the method.

One limitation which cannot be overcome is the inability to represent completely delocalized and unbound states, which is to be expected with a localized basis. With increasing NGWF radii the eigenvalues tend towards the correct Kohn-Sham eigenvalues, however when one uses such large radii the prefactor of the calculation becomes dominant, so that even though the overall behaviour is still linear-scaling, the crossover point at which the method becomes quicker than cubic-scaling codes will occur at systems with a greater number of atoms. However, for applications considered here, notably the calculation of optical absorption spectra, often only lower energy bound states are required, as many of the interesting features in optical spectra are transitions between bands close to the gap and one is interested in a relatively small energy range. Therefore in practice this limitation on the method is less serious than it first appears to be. Additionally, it has been observed that the lower energy conduction states converge with respect to NGWF radius faster than those with higher energy, and so if the lower energy bound states only are considered, it no longer becomes necessary to use such large NGWF radii to achieve a good level of convergence in the optical absorption spectra. It was for this reason that smaller conduction NGWF radii were used for the absorption spectra compared to the DOS of metal-free phthalocyanine, as presented in Sect. 7.1.

As well as the above-mentioned problems, there are a number of parameters which require more careful consideration when selecting appropriate values than in a ground state ONETEP calculation, where they can be set automatically. This includes the number of conduction states one is trying to represent, the number of NGWFs one chooses for each atom, the number of additional states to be optimized and the number of iterations for which these extra states are optimized. Some of these parameters, such as the number of iterations to perform in the first stage of the local minima avoiding scheme, have less of an effect on the final result, but for many of these parameters, the effect of different values appears to be strongly system-dependent. As discussed in Sect. 6.2.2, it is therefore vital that careful convergence tests are performed to ensure the accuracy of the final results.

It should also be noted that the iterative energy minimization scheme used here requires the presence of a band gap, which for the conduction calculation translates as a gap between the highest optimized conduction state, and the lowest unoptimized conduction state. As one approaches the continuum of conduction states, this gap will become increasingly small, which could result in poor convergence behaviour.

Finally, we observe that whilst problems have been encountered with the projection method, a clear strategy has been outlined both for identifying and resolving them. Therefore, providing one follows the systematic convergence procedures required, we expect that accurate results can be obtained for a wide range of systems.

References

1. Kokalj, A.: Comp. Mater. Sci. **28**(2), 155 (2003)
2. Ratcliff, L.E., Hine, N.D.M., Haynes, P.D.: Phys. Rev. B **84**, 165131 (2011)
3. Kobayashi, N., Nakajima, S.-i., Ogata, H., Fukuda, T.: Chem. Eur. J. **10**, 6294 (2004)
4. Shen, X., Sun, L., Benassi, E., Shen, Z., et al.: J. Chem. Phys. **132**(5), 054703 (2010)
5. Hoskins, B.F., Mason, S.A. White, J.C.B.: J. Chem. Soc. D 554b–555 (1969)
6. Fukuda, R., Ehara, M., Nakatsuji, H.: J. Chem. Phys. **133**(14), 144316 (2010)
7. Gong, X.D., Xiao, H.M., Tian, H.: Int. J. Quantum Chem. **86**, 531 (2002)
8. Cortina, H., Senent, M.L., Smeyers, Y.G.: J. Phys. Chem. A **107**(42), 8968 (2003)
9. Day, P.N., Wang, Z., Pachter, R.: J. Mol. Struct. THEOCHEM **455**(1), 33 (1998)
10. Ambrosch-Draxl, C., Majewski, J.A., Vogl, P., Leising, G.: Phys. Rev. B **51**(15), 9668 (1995)
11. Burroughes, J.H., Bradley, D.D.C., Brown, A.R., Marks, R.N., et al.: Nature **347**, 539 (1990)
12. Yu, G.: Synth. Met. **80**(2), 143 (1996)
13. Chiang, C.K., Fincher, C.R., Park, Y.W., Heeger, A.J., et al.: Phys. Rev. Lett. **39**, 1098 (1977)
14. Sariciftci, N.S., Smilowitz, L., Heeger, A.J., Wudl, F.: Science **258**(5087), 1474 (1992)
15. Heeger, A.J.: Rev. Mod. Phys. **73**(3), 681 (2001)
16. Lane, P., Liess, M., Vardeny, Z., Hamaguchi, M., et al.: Synth. Met. **84**(1–3), 641 (1997)
17. Yang, Y., Pei, Q.: Appl. Phys. Lett. **68**, 2708 (1996)
18. Moliton, A., Hiorns, R.C.: Polym. Int. **53**(10), 1397 (2004)
19. Wang, Y.: Nature **356**, 585 (1992)
20. Badamshina, E., Gafurova, M.: Polym. Sci. Ser. B Polym. Chem. **50**, 215 (2008)
21. Gomes da Costa, P., Dandrea, R.G., Conwell, E.M.: Phys. Rev. B **47**, 1800 (1993)

Chapter 8
Conclusion

We began this thesis by observing that the field of experimental spectroscopy has made a significant impact on our knowledge of the properties of materials, in particular relating to their electronic structures. However, in order to fully understand this information, the ability to theoretically determine spectra is a vital tool. We have therefore presented a method of calculating optical absorption spectra within a linear-scaling DFT framework, with the twin aims of allowing ab initio theoretical spectroscopy calculations on large systems and providing a means of extracting useful information from large scale simulations of materials which can be directly compared to experiment. The central component of this approach was the development of a method for accurately calculating the unoccupied Kohn-Sham states, which are not well represented within the optimized local orbital basis sets used for ground state linear-scaling DFT calculations.

In order to more accurately represent the unoccupied states, a method was developed for optimizing a second set of local orbitals in a non self-consistent calculation. Three different methods for doing so were compared within a toy model. The most efficient method involved the use of the density operator to project out all contributions of the occupied states to the total energy, so that the new set of conduction orbitals could be minimized with respect to the total energy of a selected number of conduction states. This projection method was implemented with the linear-scaling DFT code ONETEP and excellent agreement was demonstrated with plane-wave DFT results.

In addition, we have also explored some of the features required for an accurate localized basis set and the impact of less than ideal basis sets on the generation of band structures. Two different methods for generating band structures in localized basis sets were compared, which were shown to be equivalent for a good, fully converged basis set, as would be expected. However, for the simple nearest neighbour basis sets considered, considerable differences between the two methods were observed, proving both that a nearest neighbour approach is too simplistic for the accurate calculation of one-dimensional band structures and that one of the two methods is more appropriate for use in ONETEP. Furthermore, the deficiencies of these simple

L. Ratcliff, *Optical Absorption Spectra Calculated Using Linear-Scaling Density-Functional Theory*, Springer Theses, DOI: 10.1007/978-3-319-00339-9_8,
© Springer International Publishing Switzerland 2013

basis sets allowed us to identify some of the causes responsible for inaccuracies in band structures calculated in ONETEP.

The existence of a localized basis set capable of representing the Kohn-Sham conduction states in ONETEP enabled us to implement the calculation of optical absorption spectra using Fermi's golden rule. Whilst other methods could be used to calculate absorption spectra to a greater degree of accuracy, they are more expensive to implement and so prohibit access to the large systems that one aims to study with a linear-scaling DFT method. The optical absorption spectra calculated in ONETEP also showed excellent agreement with traditional plane-wave results.

Some problems were encountered with local minima when optimizing these conduction orbitals, and so a strategy was developed for both identifying and avoiding such problems. Whilst there are some limitations to the use of localized basis functions, in that they are not ideal for the representation of higher-energy delocalized conduction states, such higher energy states are not necessary for the calculation of optical spectra within the energy windows considered in this work.

Through application to both molecular and extended systems, we have demonstrated the ability of our method to identify the transitions responsible for a given peak in a spectra, and differentiate between results from very similar atomic structures. Finally, we calculated the absorption spectra for a doped conjugated polymer, which could be a candidate for use in photovoltaic applications, demonstrating the possibility of using our method for applications of technological importance.

8.1 Future Work

There are three main ways in which the work presented in this thesis could be further developed in future. Firstly, the calculation of optical absorption spectra could be further developed to include a sum over states at a number of different $\mathbf{k}$-points, rather than just at the Γ-point. This could be achieved using either the $\mathbf{k} \cdot \mathbf{p}$ or tight-binding style methods for band structure interpolation.

The method we have presented here could also be extended to the calculation of other types of spectra. This could include the calculation of EELS and XAS, which involve transitions between core and unoccupied states. This would therefore require some additional effects to be taken into account, including, for example, core hole screening.

Finally, whilst our method is an important first step in calculating optical absorption spectra, there are a number of approximations involved. In order to achieve truly excellent agreement with experiment, it will be necessary in future to apply methods which allow for the calculation of more accurate results. As we have previously mentioned, recent research has been done on decreasing the computational costs associated with such methods, including the use of the GW approximation combined with the Bethe-Salpeter equation. It is anticipated that such progress will continue, and it will therefore be necessary to implement a similar approach within a localized orbital framework in the future.